AF560294

Innovations of University Extension in Indian Agriculture

NIPA® GENX ELECTRONIC RESOURCES & SOLUTIONS P. LTD.
New Delhi-110034

About the Authors

Dr. R Venkattakumar, He has been serving in ICAR for past 25 years and presently serving as Head, Extension Systems Management at National Academy of Agricultural Research Management (NAARM), Hyderabad. He received 16 awards and recognitions. His major areas of interest include Mass and Business Communication, Adoption and Diffusion of Farm Innovations, Impact Assessment, Farmer Producer Organizations (FPOs), Rural Entrepreneurship, and Farmer Groups. He published more than 100 publications, in the field of Agricultural Extension. He received specialized training from Michigan State University (MSU) of USA towards promoting entrepreneurship in agriculture.

Dr. P Venkatesan, Principal Scientist at ICAR-NAARM, Hyderabad, is a leading expert in agricultural extension and research management. With over 27 years of experience, he has led 19 completed and 4 ongoing national projects, aligning research with farmers' needs and national priorities. He has conducted 67 training programs, mentored two Ph.Ds. & 20 postgraduates, and developed innovations like the Agri-Prioritize App and ADOPT tool. A prolific author with 57 research articles and 8 books, his work spans around capacity building, policy shaping, and digital knowledge systems. His contributions have significantly impacted rural livelihoods and the future of agricultural education and research in India.

Dr. Ch Srinivasa Rao, He is currently serving as Director and Vice Chancellor of ICAR-Indian Agricultural Research Institute (IARI), New Delhi. He served as Project Coordinator, All India Coordinated Research Project on Dryland Agriculture (AICRPDA), Hyderabad and several other positions at Indian Institute of Soil Science (IISS), Bhopal; Indian Institute of Pulses Research (IIPR), Kanpur; CRIDA, Hyderabad, ICRISAT, Patancheru, Hyderabad and NAARM, Hyderabad. He has 32 years of experience in research management. He is the recipient of 46 National and International awards. He is a Fellow of several reputed professional bodies. He is credited with more than 640 publications.

Dr. Gopal Lal, As a Horticulturist, he served in RAU, Rajasthan; ICAR-CAZRI, Jodhpur; CAU, Pasighat; ICAR-NRCSS, Ajmer and Cauvery Water Management Authority, New Delhi. He has more than 29 years of experience in NARS. Presently he is serving as Acting Director, ICAR-NAARM, Hyderabad. His areas of interest include Horticulture, MIS, Organic Farming, Protected Cultivation, PHT and Value addition, Value Chain Management and Natural Resources Management. He has guided 13 Students for M.Sc. and Ph.D. Degree. He has developed 13 Varieties of Seed Spices, 60 Technologies and 35 Value added products and published more than 360 publications, and recipient of 7 fellowships.

Innovations of University Extension in Indian Agriculture

R Venkattakumar
Head
Extension Systems Management
National Academy of Agricultural Research Management (NAARM)
Hyderabad, Telangana - 500 030

P Venkatesan
Principal Scientist
ICAR-NAARM
Hyderabad, Telangana - 500 030

Ch Srinivasa Rao
Director and Vice Chancellor
ICAR-Indian Agricultural Research Institute (IARI)
New Delhi - 110 012

Gopal Lal
Acting Director
ICAR-NAARM
Hyderabad, Telangana - 500 030

NIPA® GENX ELECTRONIC RESOURCES & SOLUTIONS P. LTD.
New Delhi-110034

NIPA® GENX ELECTRONIC RESOURCES & SOLUTIONS P. LTD.

101,103, Vikas Surya Plaza, CU Block
L.S.C. Market, Pitam Pura, New Delhi-110034
Ph : +91-11-43860225, Mob.: +91 9717133558, 9540816132
E-mail: newindiapublishingagency@gmail.com
Website: www.nipaersources.com

Editorial support

Smt. G. Aneeja (Chief Technical Officer, ICAR-NAARM, Hyderabad)

Secretarial assistance:

D. Bhavana (YP-II, XSM Division, ICAR-NAARM, Hyderabad)

B. Rajitha (YP-II, XSM Division, ICAR-NAARM, Hyderabad)

Acknowledgement

The authors are highly grateful to ICAR-NAARM, Hyderabad for providing all necessary infrastructure for publishing this book.

Print ISBN: 978-93-72197-72-3

ebook ISBN: 978-93-72197-16-7

Composed and Designed by NIPA®.

डॉ. एम. एल. जाट
सचिव (डेयर) एवं महानिदेशक (भाकृअनुप)

Dr M. L. Jat
SECRETARY (DARE) & DIRECTOR GENERAL (ICAR)

भारत सरकार
कृषि अनुसंधान और शिक्षा विभाग एवं
भारतीय कृषि अनुसंधान परिषद
कृषि एवं किसान कल्याण मंत्रालय, कृषि भवन, नई दिल्ली 110 001

GOVERNMENT OF INDIA
DEPARTMENT OF AGRICULTURAL RESEARCH & EDUCATION (DARE)
AND
INDIAN COUNCIL OF AGRICULTURAL RESEARCH (ICAR)
MINISTRY OF AGRICULTURE AND FARMERS WELFARE
KRISHI BHAVAN, NEW DELHI 110 001
Tel.: 23382629; 23386711 Fax: 91-11-23384773
E-mail: dg.icar@nic.in

FOREWORD

The National Agricultural Extension System (NAES) of the country engaged several need-based and situation-specific modes, methodologies and approaches to transfer advanced agricultural innovations to the farming community. Technology transfer to the end users is very vital for achieving inclusive growth in agriculture. "New Age Farmers" need technology-based entrepreneurship promotion for which roles and responsibilities of NAES are crucial.

Directorate of Extension Education (DEEs) play an important role in disseminating regional-specific technologies, developed by State Agricultural Universities (SAUs). I am happy to note that these technologies have been compiled in the form of a book entitled "Innovations of University Extension in India".

I compliment the editors of the book and hope that it will serve as a valuable reference material among the extension functionaries and practitioners.

(M. L. Jat)

Dated the 8th July, 2025
New Delhi

ICAR-National Academy of Agricultural Research Management (NAARM)

Rajendranagar, Hyderabad-500030

Dated: 27.10.2025

Preface

The National Agricultural Research System (NARS) strive to develop innovations that suits the emerging demands of stakeholders including farmers. National Agricultural Extension System (NAES) has been disseminating the innovations developed by NARS, so that these innovations can be applied across technology utilization ecosystems of the country.

State Agricultural Universities (SAUs) of the NARS indulge in developing regional-specific innovations. These innovations are ably disseminated to the stakeholders through Directorates of Extension Education (DEE) of SAUs. DEEs employ innovative extension methodologies and approaches, which are location-specific and demand-drive, However, these approaches are unique to the respective DEEs.

The methodologies and approaches adopted by DEEs include personal, group and mass modes of technology dissemination. They also engage innovative information and communication technologies (ICTs) towards communication of innovations. Apart from technology dissemination, they also involve themselves in promotion of technology-based entrepreneurship, so that business opportunities in agriculture and rural landscape can be enhanced thereby address the unemployment issues in rural India.

Cross-learning is one of the crucial learning approaches, which facilitates studying the proven approaches of different situations and test-verifying in the situations of the learners. By the way, cross learning about innovative extension approaches adopted by DEEs may enable applications of proven extension methodologies in newer situations.

The very purpose of this book is to capture innovative extension methodologies and approaches adopted by SAUs of the country, so that it will be a valuable reference material and source for cross-learning among the DEES. We, the authors, take this opportunity to thank all the contributors of 24 chapters representing 15 DEEs for sharing information about innovative extension approaches.

R Venkattakumar
P Venkatesan
Ch Srinivasa Rao
Gopal Lal

Contents

1

Fostering Sustainable Agriculture: The Leadership of Directorate of Extension, Kerala Agricultural University

***Jacob John*[1] *and V.M. Hima*[2]**

[1]Director of Extension and [2]Assistant Professor and Scientific Officer to Director of Extension, Kerala Agricultural University, Kerala

Email: de@kau.in

Abstract

The Directorate of Extension (DoE) at Kerala Agricultural University (KAU) plays a vital role in technology transfer and capacity building in agriculture and allied sectors. Collaborating with the development departments of Kerala State, the DoE ensures the effective dissemination of new technologies, trends, and research findings. Utilizing multiple platforms particularly, Krishi Vigyan Kendras (KVKs), Communication Centre, Central Training Institute (CTI) and Agricultural Technology and Information Centre (ATIC), the directorate strives to translate the technological advancements of KAU into sustainable development strategies. Through targeted interventions, effective stakeholder interfaces, and advanced management practices, the DoE contributes to food sovereignty and livelihood security across Kerala. This chapter outlines the initiatives, reach, output, and impact of Directorate of Extension, KAU, while providing strategic recommendations for strengthening extension management.

Keywords: *Technology transfer, capacity building, sustainable agriculture, extension management, Krishi Vigyan Kendras, Kerala Agricultural University, food sovereignty, livelihood security*

Introduction

Technology transfer and capacity building are critical to the advancement of agriculture, particularly in ensuring that research findings and innovations reach stakeholders efficiently. The Directorate of Extension (DoE) at Kerala Agricultural University (KAU) is a key player in this process, liaising with state development departments to drive technology dissemination in the agriculture and allied sectors. The DoE collaborates closely with development departments of agri and allied areas in the State, acting as a vital link to ensure that agricultural advancements reach the grassroots level. The Directorate also integrates institutional mechanisms to support this technology transfer. Through a network of Krishi Vigyan Kendras (KVKs) and several institutional mechanisms, the Directorate integrates new agricultural technologies into mainstream practices, focusing on both skill development and sustainable development strategies. It conducts workshops, seminars, and training programs focused on enhancing the skills of extension workers, farmers, and other stakeholders. The focus is not just on disseminating new technologies but also on building long-term capacity through skill development. This ensures that the stakeholders are well-equipped to implement the innovations in a way that promotes efficiency, productivity, and sustainability.

In essence, the DoE at KAU is a driving force in modernizing agriculture in Kerala. Through its structured efforts in technology transfer, skill enhancement, and the promotion of sustainable practices, it enables the agricultural sector to keep pace with scientific advancements while ensuring that the benefits are accessible and applicable at the field level.

Initiatives / Interventions of DoE, KAU

The Directorate of Extension is crucial in bridging the gap between agricultural research and field-level application through various innovative initiatives. Key interventions include the popularization of agricultural technologies, primarily through farm-based production programs, Block Level Agricultural Knowledge Centres (BLAKC), and the *Mukhamukham* (face-to-face) Agro Clinic. The Directorate also focuses on capacity-building programs, entrepreneurship development, and ICT-enabled technology dissemination. Additionally, the Directorate supports vulnerable sections of society, develops Farmer Producer Organizations (FPOs), and implements state-funded development projects, contributing to the holistic growth of Kerala's agricultural sector.

I. Popularization of Agricultural Technologies

- **Farm Plan based Production Program:** The Kerala State Government, based on the background field level information given by the KAU, has embarked upon a flagship program with farm plan-based approach as a strategy for holistic development of individual farms and to improve the income of the famer and provide livelihood security on long term perspective. This is contrary to the crop-centric approach adopted earlier. On behalf of Kerala Agricultural University, Directorate of Extension was entrusted, with adequate funds, for providing technical support for the implementation of the scheme by way of creating master trainers, preparation and implementation of a technical module and extension activities for disseminating technical inputs on a regular basis by conducting front line demonstrations, on farm development of technology, workshops, farmer-extension-scientists interactions *etc.* in all the 152 blocks of the State.

- **Block Level Agricultural Knowledge Centres (BLAKC):** Kerala Agricultural University (KAU) has launched a pioneering initiative in establishing Block Level Agricultural Knowledge Centres (BLAKCs) across all the 152 blocks in Kerala. These AKCs serve as key block-level entities aimed at disseminating cutting-edge knowledge in agriculture and allied sectors to farmers and development personnel, thereby supporting agricultural production at the grassroots level. Under this innovative model, each AKC is linked to one agricultural scientist.

To address agriculture-related challenges specific to each block, scientists from KAU, who serve as Nodal Officers of these Block Level Agricultural Knowledge Centres (BLAKCs), conducted Front Line Demonstrations (FLDs) of KAU technologies in all 152 blocks. In some areas, multiple FLDs were carried out in collaboration with the State Department of Agriculture.

Additionally, the Directorate of Extension (DoE) successfully facilitated the smooth implementation of the Krishi Bhavan Attachment program (KBA 1201 course), "Participatory Learning on Farming Situation and Extension Interventions," for undergraduate students of Kerala Agricultural University, leveraging the expertise of the Nodal Officers of the BLAKCs.

II. Interface Initiatives

KAU has launched '*Mukhamukam* (Agro Clinic & Farm Advisory Conclave)', a unique interface initiative that allows farmers to communicate directly with an interdisciplinary team of scientists. KAU, in collaboration with the Agriculture Technology Management Agency (ATMA), Kerala is implementing the *Mukhamukham* scheme to identify lead farmers. Lead farmers visit the interdisciplinary team, and solutions are recommended.

- **Face-to-face Agricultural Science Debate Program** has been held continuously for the last three years to disseminate agricultural knowledge in the general society and hold debates. Trainees from all over Kerala participate in this program, which is very useful for public activists and people.
- **Workshops / Seminars:** To strengthen the public interface of the University, the Directorate of Extension organized several workshops, seminars, and interactive sessions.

- **Farm Advisory and Diagnostic Services:** Farm Advisory and Diagnostic Services were offered through field visits, direct interaction, phone calls, and e-mails. To tackle the field difficulties and sudden outbreaks of pests and diseases, DoE formulated Multi-Disciplinary Diagnostic Team (MDDT), containing multidisciplinary teams of scientists from Kerala Agricultural University.
- **Monthly Technology Advisory (MTA)** meetings in collaboration with ATMA are being held under the aegis of DoE, KAU to extension personnels from the State Department of Agriculture, ATMA and VFPCK (Vegetable and Fruit Promotion Council Kerala).

III. Capacity Building Programs

Under the guidance of DoE, several capacity-building programs have been organized by the Central Training Institute, Communication Centre, and Krishi Vigyan Kendras, independently and in association with agri and allied departments. In this context, tailored training programs were developed and implemented for extension personnel, rural youth, women farmers, farmers, entrepreneurs, and school students. Notably, the DAESI and INM programs stand out as significant initiatives.

- **Diploma Programme for Agri-Input Dealers (DAESI)**: In collaboration with MANAGE and ATMA, DoE is implementing this program which was intended to transform input dealers into para-extension professionals capable of addressing day-to-day problems faced by the farmers at the field level, through various KVKs and other Institutes of KAU.
- **Integrated Nutrient Management (INM) course**: DoE Implemented the INM course program that was introduced by the Central Government for dealers of fertilizer and pesticides. The Central Training Institute offers training courses in agriculture and related fields through a specialized framework known as the School of Entrepreneurship Development.

IV. Education Programs

- **Introduction of Academic Programs**: Under the patronage of DoE, KAU a Ph.D. program in Animal Science and a Post Graduate Diploma in Integrated Farm Management was launched through the Communication Centre, thereby enhancing academic offerings. (Details are available in the site https://www.kau.in)
- **Certificate Courses:** To address the emerging need for trained professionals in agriculture sector, the Directorate of Extension, KAU launched forty-one, self-supporting, new generation certificate courses

through the various colleges and stations spread across Kerala. (Details are available in the site https://www.cti.kau.in)

- **Massive Open Online Courses:** The Massive Open Online Courses (MOOCs) for farmers, extensionists, students, farm entrepreneur groups and other agri-stake holders were organized by the Centre for e-learning (CeL, KAU) under the aegis of Directorate of Extension. (Details are available in the site https://www.celkau.in)
- **Production of Educational Videos**: Initiated a high-quality lecture series utilizing advanced digital technologies, graphics, text animations, and stock footage to create engaging educational content.
- **Comprehensive Student Support:** Ensured effective course management and provided guidance to students through dedicated mentorship by scientists within the Directorate of Extension. The Directorate of Extension is set to enhance the admission processes for courses offered by Kerala Agricultural University (KAU) during 2024-25 by organizing a statewide campaign to promote new-generation courses through various audio-visual media and channels, besides conducting one-day educational awareness seminars titled "*Mikacha Thozhilunu Karshika Coursukal*" at six selected regions of Kerala.

V. Dissemination of Information through Publications and Mass Media

Extensive coverage on innovative crop production and value addition practices was provided through mass media. The latest developments in the agricultural sector were shared for broadcast on All India Radio (paid contract between DoE and AIR for an entire year) and Doordarshan. Scientists from KAU participated as resource persons in the *'Farm and Home'* program on All India Radio and other channels. The Public Relations Unit under the Directorate of Extension managed the media connections and coordination activities of the university. Additionally, scientific advisory services were offered through the Farm Information Bureau.

- The KAU semi-scientific farm journal *Kalpadhenu* (with ISSN since 2024), a quarterly, extends information to farmers and the scientific community, with each issue focused on a major issue in the farming sector.

  (Details are available in the site https://www.ccmannuthy.kau.in)
- DoE released the '*Package of Practices Recommendations*: *Crops 2024'* and obtained International Standard Book Number (ISBN). The "*Package of Practices Recommendations: Crops 2024*" is a detailed guide published by Kerala Agricultural University. It provides

scientifically developed, region-specific guidelines on best practices for cultivating various crops, managing soil health, disease and pest control, and improving yields. This publication is crucial for farmers and agricultural extension workers as it helps bridge the gap between research and practical application in the field. The Malayalam version of the same is now ready for release.

Additionally, the publication of '*Organic Package of Practices*' is indeed a notable initiative by the Director of Extension. This comprehensive resource offers detailed guidance on organic production methods tailored to a wide array of crops across diverse categories, supporting sustainable agriculture and empowering farmers with effective organic practices. It serves as an essential reference for practitioners, agricultural advisors, and extension workers in promoting eco-friendly and sustainable crop management in Kerala.

- KAU News: A publication issued by the Kerala Agricultural University (KAU). It serves as a medium to disseminate information related to the academic, research and extension activities of the university. The newsletter typically highlights the latest developments, breakthroughs, events, and initiatives undertaken by KAU. Its periodicity has been enhanced from quarterly to bimonthly from January 2024. (Available at https://www.extension.kau.in)
- Launched online sale of KAU publications brought out by DoE, KAU with '*Keralagro*' brand name through Amazon. All the publications have ISBN.
- Released coffee table book "*Abode*", describing the activities of various units under DoE, KAU in January 2024.

VI. ICT-enabled Technology Development and Dissemination

The Directorate of Extension, KAU produced videos on various centers of KAU and short videos on technological advancements under the leadership of ATIC (Agricultural Technology Information Centre), and the same have been uploaded in KAU websites, you tube channel, face book page, *etc.,* for ensuring wider dissemination. Over 200 videos uploaded on various social media, including approximately 150 focussing on agricultural technologies evolved from KAU, has attracted more than 20,000 viewers per video. Some other initiatives include :

- Established an advanced sound proof, recording cum editing studio at ATIC, Mannuthy.

- Developed a mobile application named *Farm Extension Manager* [FEM@Mobile] rated as the best application in Kerala, with over 50,000 downloads and a 4.5 rating. It covers specially categorised information on more than 100 crops. It is now available on Playstore.

VII. Entrepreneurship Development in Agriculture

KAU, under the guidance of the Director of Extension, supports Agri-Business Incubation (ABI) Centres. These centres provide entrepreneurs with the necessary infrastructure, mentoring, and technical support. The centres act as a platform for farmers and agri-startups to develop and commercialize their innovative ideas.

- **KAU RKVY-RAFTAAR-Agribusiness Incubator (KAU RABI):** To rejuvenate and cater the needs and modalities for agribusiness promotion, the Government of India introduced a revamped "Rashtriya Krishi Vikas Yojana – Remunerative Approaches for Agriculture and Allied Sector Rejuvenation" (RKVY-RAFTAAR) under the Ministry of Agriculture and Farmers' Welfare (MoA& FW). It is aimed at strengthening infrastructure in Agriculture and Allied sectors to promote Agripreneurship and Agribusiness by facilitating financial aid and nurturing a system of business incubation. The Kerala Agricultural University established KAU RAFTAAR Agri Business Incubation (KAU- RABI) Centre in 2019 to take forward the vision and objectives of RKVY RAFTAAR. KAU RABI envisages agribusiness incubation by helping creative minds to innovate and audaciously use technologies for venture creation in agriculture. KAU RABI supports agribusiness incubation by tapping innovations and technologies for venture creation in agriculture. RAFTAAR Agri Business Incubator conducted several entrepreneurship development training programs to entrepreneurs, farmers, SHG, FPO's *etc.* To promote innovation, entrepreneurship and business creation in agriculture and allied sector, Agri Business Incubator has successfully incubated and assisted 245 innovative budding startups through its five cohorts of 8 weeks RKVY-RAFTAAR Innovation and Agripreneurship development programs. A total of 84 startups have been selected in various areas of agriculture and allied sectors under this program for providing pre-seed stage and seed-stage funding support of an amount of 10.97 crore by Ministry of Agriculture and Farmers Welfare. RAFTAAR - RABI has now reached its sixth Cohort.
- **Agri-Clinics and Agri-Business Centres (ACABCs):** Two Agri-Clinics and Agri-Business Centres (ACABCs) are operational at the Training Service Scheme in Vellayani and RARS, Pattambi, in collaboration

with MANAGE, Hyderabad, aimed at enhancing the employability of agricultural graduates. To encourage self-employment among these graduates, numerous candidates received agripreneurship training over the past financial year. An Agri-clinic was also established at the station by the Extension Training Centre (ETC), Manjeswar, with the goal of diagnosing pests and diseases and providing recommendations for Integrated Pest Management (IPM) strategies.

DoE is fostering innovation and entrepreneurship through the creation of additional business incubation centres, for providing a conducive environment for the development of agri-businesses.

- **Farmer Producer Organizations (FPO):** By promoting Farmer Producer Organizations (FPOs) and collective marketing models, the Director of Extension encourages farmers to come together as entrepreneurial units. FPOs help farmers gain better market access, improve their bargaining power, and explore value chain opportunities. The Directorate of Extension through the various KVKs was instrumental in forming and hand holding FPOs across the state with the involvement of KVKs. In 2024, the Mayyil Rice Producer Company, an FPO established by KVK Kannurunder DoE, KAU, received the state award for excellence as the top-performing FPO/FPC in agriculture. Recognized for its outstanding achievements, Mayyil Rice Producer Company has also been selected as one among the 20 Best FPOs across India.

- **Satellite Incubation Centre (SIC):** ICAR-CTCRI in collaboration with DoE, KAU has proposed to establish a Satellite Incubation Centre (SIC) at Regional Agricultural Research Station, KAU, Pattambi. The proposed centre will focus on developing sustainable value chains of biofortified sweet potato in Attapadi (Tribal area) and other tribal areas of Kerala using the vast network of tuber crops farmers, traders, startups/ MSMEs associated with ICAR-CTCRI.

VIII. Sale of Quality Inputs

The "*Vithu Vandi*" (Seeds on wheels") is a program introduced with this intention, during the period of the pandemic. It helped many farmers to get the seeds and planting materials at their doorstep and raise crops during the lockdown period. The service of Mobile Exhibition Unit/Vithu vandi is still being used for the sale of seeds.

Distribution of quality seeds, seedlings, inputs and value-added products are carried out through the Agricultural Technology Information Centre, Communication Centre, Agricultural Information and Sales Centre and KVKs.

IX. Extension Support to the Vulnerable and Marginalized Sections of the Society

- **Schedule Castes Sub Plan (SC/SP) and Tribal Sub Plan (TSP) Programmes:** Under the ICAR Development Grant - sub component "Scheduled Caste Sub Plan", a project, is being implemented through all the KVKs under the control of DoE aiming to conduct trainings, counselling and upliftment programs for the beneficiary group under SC community.
- **Extension Centre for Tribal Areas:** Utilizing the State Plan funds of DoE, the KVK, Palakkad is implementing this project for the benefit of tribes in Palakkad. Targeted strategies for the benefit of SC-ST farmers have been implemented in areas such as millet production, planting material distribution, technology dissemination *etc.*

X. Implementation of National Initiatives (under leadership of ICAR-ATARI-Bengaluru through DoE, KAU)

- **National Innovations on Climate Resilient Agriculture (NICRA):** This project, launched by the Indian Council of Agricultural Research (ICAR) under the Ministry of Agriculture and Farmers Welfare in 2011, focuses on research and technology demonstration to help farmers adapt to and mitigate climate change impacts on agriculture. NICRA is implemented in Kerala through KAU Krishi Vigyan Kendras (KVKs) in Kottayam, Wayanad, Kannur, and Palakkad districts
- **District Agromet Unit (DAMU)**: DAMU established under the Gramin Krishi Mausam Seva (GKMS) scheme by the Ministry of Earth Sciences (MoES) in partnership with ICAR and IMD, offers weather-based advisory services to farmers through KAU KVKs in Kollam, Malappuram, and Palakkad, publishing the "*Meghasandhesa*" e-newsletter.
- **Attracting and Retaining Youth in Agriculture (ARYA)**: The project, initiated in 2015-16, aims to engage rural youth in agriculture by fostering entrepreneurship in agricultural and allied sectors. KVKs in Kannur and Malappuram are involved, providing training to youth in areas such as coconut, jackfruit, and vegetable processing to support income generation and rural development.

XI. Major Development Programmes through Plan/externally Aided Projects

The Directorate of Extension implemented and monitored extension plan projects funded by the State Government. The nature of extension plan projects ranged from HRD programs for agriculture stakeholders to infrastructure development for strengthening the transfer of technology system of Kerala Agricultural University.

Table 1. Major Extension Projects (2023-24)

S. No.	Name of Project	Funding Agency	Financial outlay (Rs. Lakhs)
1	Strengthening Extension and Outreach Activities	Government of Kerala	30.0
2	Training programs by research stations	Government of Kerala	5.0
3	NABARD "Promotion of Integrated farming thorough skill upgradation of master farmers"	NABARD	24.0
4	Documenting cult, culture and agricultural relations in Kerala	State Plan	10.0
5	Maintenance of Kissan Eco park & strengthening of common facilitation centre at Communication Centre, Mannuthy	State Plan	5.0
6	Strengthening technology dissemination through technology micro videos	State Plan	5.0
7	Establishment of exotic fruit plants demonstration unit at ETC Manjeshwar	State Plan	5.0
8	Establishment of "The State Coconut Museum" at Coconut Research Station, Balaramapuram	State Plan	10.0
9	Establishment of a Tuber crop museum at KVK, Thrissur	State Plan	2.0
10	Establishment of Common Facility Centre at KVK Thrissur	NABARD	17.685
11	Establishing of a mass production unit for bio control agents at KVK, Kannur	State Horticulture Mission	44.775
12	Network project on Trichogramma mass production (5 centres)	State Plan	10.0
13	Agri innovation cum incubation centre at CoA, Vellayani	State Plan	10.0

XII. Other Critical Interventions

- KAU is planning to move more towards developing protocols for different crops, to increase the production and productivity of crops. In the protocol based advisory services, the entire farming operations will be linked with the date of planting and plant growth. So, the farmer can get correct advice on when to go for various operations. In the new model crowns will be given the correct nutrient and appropriate dosage based on soil test results also. In this context, DoE released scientific protocol for paddy crop of Kole land.

- The GIS mapping of Kole lands and preparation of an Atlas of Kole lands of Kerala with padashekarams for crop planning is another activity of DoE, KAU which will greatly influence the paddy cultivation in our state.
- DoE has entered into new collaborations and is continuing existing ones with the Kerala State Department of Agriculture, MANAGE, CIFT, NABARD, Directorate of Cashewnut and Cocoa Development (DCCD), LSGD, and the Kerala State Department of VHSE, to implement various programs.
- KAU-FPO Linkage Project (KFL): A remarkable initiative of the Directorate of Extension, which aims to empower Farmer Producer Organizations (FPOs) in Kerala through a strategic partnership with KAU. This initiative recognizes the transformative potential of FPOs in the agricultural sector and seeks to bridge the gap between academic expertise and practical experience to promote sustainable growth. Through collaborative efforts, the KFL project aims to enhance profitability and support the revitalization of agriculture in the state. The Kerala State Government has sanctioned this project with funds of RKVY through SFAC (Small Farmers Agribusiness Consortium).
- Under the leadership of Central Training Institute, DoE hosted a week-long program on "*New Competencies, Career Opportunities, and Research Priorities in Agricultural Extension*", National Young Professional Development Program (NYPDP) in collaboration with the National Institute of Agricultural Extension Management (MANAGE), Hyderabad. This MANAGE sponsored program aimed to orient participants on emerging areas in Agricultural Extension research, practice, and policy. KAU's commitment to nurturing innovative thinking and MANAGE's expertise will empower the next generation of agricultural extension leaders.
- DoE, KAU in collaboration with CAB International (CABI) South Asia and the M.S. Swaminathan Research Foundation (MSSRF), successfully organized a four-day training program under the leadership of CTI, for Plant Doctors and Master Trainers that aims to enhance the skills of agricultural extension workers in diagnosing and managing crop pests and diseases. This initiative emphasizes the importance of agricultural resilience in Kerala and promotes Integrated Pest Management (IPM)

practices, thereby contributing to improved food security and the livelihoods of farmers.

- Established a common facilitation centre at Communication centre, Mannuthy
- In commemoration of the Golden Jubilee of Kerala Agricultural University, a memoir (KAU Golden Jubilee Souvenir) that highlights the University's history, achievements and the contributions of KAU institutions over 50 years along with messages of eminent personalities and erstwhile Vice-Chancellors and the memories of selected KAU alumni were compiled under the leadership of Central Training Institute, DoE, Mannuthy.

XIII. Supporting Units of Directorate of Extension, KAU

- **Agricultural Technology and Information Centre (ATIC):** ATIC, guided by the DoE, acts as a resource hub for the public to access agricultural technologies and expert advice. The centre has also produced videos on various university centres and agricultural techniques to support knowledge transfer.
- **Communication Centre:** The DoE oversees the dissemination of agricultural information, research trends, and findings through publications and media to ensure widespread awareness among farmers and stakeholders.
- **Central Training Institute (CTI):** Under the leadership of the DoE, the CTI conducts various training programs aimed at skill development and capacity building in agriculture and related fields.
- **Krishi Vigyan Kendras (KVKs):** The DoE manages seven KVKs under KAU across the state (Kannur, Wayanad, Malappuram, Thrissur, Palakkad, Kottayam, and Kollam districts) focusing on field-level demonstrations, farmer training, and outreach activities to ensure the practical application of scientific innovations. They play a pivotal role in the implementation of schemes of the Directorate of Extension, national initiatives in particular, and various central ICAR schemes.

Under the leadership of the Director of Extension, Kerala Agricultural University (KAU) successfully hosted the *Zonal Women Agri-preneur Conclave* during 2024 at its Thrissur campus, uniting KVKs from Kerala and Karnataka to promote women-led agricultural entrepreneurship, with support from ICAR and inaugurated by the Union Minister of State for Agriculture and Farmers' Welfare. Additionally, the Directorate of Extension facilitated the Annual Review and Action Plan Meeting for KVKs in Kerala and Lakshadweep at KAU Headquarters, which included the release of ICAR publications and a video produced by the Directorate of Extension.

- **Public Relations Office:** The DoE has the Public Relations Office as one of its units for publicizing technologies evolved by KAU and activities undertaken for the benefit of the farming community to the public and key stakeholders, thereby enhancing the university's visibility and impact.

The other units include Extension Training Centre, Manjeswaram, Kasaragod, Centre of Agricultural Innovations and Technology Transfer (CAITT), Vellayani, Thiruvananthapuram; Centre for e-Learning (CeL), Vellanikkara, Thrissur; Agricultural Information and Sales Centre, Kozhikode; RAFTAAR Agri-Business Incubator, Vellanikkara, Thrissur; Tribal Extension Centre, Attapady, Palakkad; and KAU Press. These initiatives reflect the DoE's commitment to effective extension management, ensuring the transfer of knowledge, technology, and skills to the agricultural community.

XIV. Reach, Output and Impact of Directorate of Extension, KAU

- **Reach**: The Director of Extension (DoE) operates extensively throughout Kerala, leveraging seven Krishi Vigyan Kendras (KVKs) as the key outreach centers for connecting with farmers at the grassroots level. These KVKs facilitate on-the-ground demonstrations, trainings, and direct farmer interactions. Additionally, the Communication Centre and Agricultural Technology and Information Centre (ATIC) play vital roles in ensuring the widespread dissemination of research findings and technological advancements. Through various media, publications, and digital platforms, these centers reach a broad audience, including both rural farmers and urban stakeholders, ensuring that KAU's innovations and knowledge impact the entire agricultural community.
- **Output**: The Directorate's major outputs include regular publications like the Package of Practices Recommendations, training sessions conducted by CTI, and the wide-scale dissemination of agricultural technologies through the KVKs. Books and booklets, along with quarterly magazines and newsletters, have been made available to the farming community.
- **Outcome / Impact**
    - Improved Crop Yields and Productivity: The adoption of advanced agricultural technologies and modern farming practices has led to significant improvements in crop yields and resource use efficiency across Kerala.

- Bridged the Research-Practice Gap: The DoE has reduced the gap between research and field-level application by transferring cutting-edge research findings directly to farmers through publications, training programs, and KVK outreach.
- Promotion of Sustainable Agricultural Practices: Sustainable methods such as organic farming, Integrated Pest Management (IPM), and eco-friendly technologies have been widely promoted, leading to improved soil health, water conservation, and environmental sustainability.
- Scaling up of agricultural technologies/models evolved from KAU on a large scale throughout the State in collaboration with the State Department of Agriculture and Farmers' Welfare utilizing their funds as per the annual plans approved by the Kerala State Planning Board for the 14th five-year plan (2022-27).
- Food Sovereignty and Livelihood Security: Increased local food production and reduced dependence on external sources have bolstered food sovereignty, while improved farming practices have provided greater livelihood security for Kerala's agricultural population.
- Empowerment of Farmers and Agripreneurs: The DoE's training programs and support for Farmer Producer Organizations (FPOs) have empowered farmers to become successful agripreneurs with improved incomes and market access (including export).
- Inclusive Reach Across Rural and Urban Areas: The DoE ensures broad dissemination of agricultural knowledge and technologies, reaching both rural farmers and urban stakeholders through KVKs, ATIC, and the Communication Centre.

These outcomes reflect the DoE's significant impact on agricultural development, sustainability, and farmer empowerment in Kerala.

Conclusion

The Directorate of Extension of KAU plays a crucial role in transferring agricultural technologies to stakeholders across Kerala. Through its extensive network of KVKs/other vibrant units and partnerships with state development departments, the DoE has successfully bridged the gap between research and practice. Its contributions to capacity building, food sovereignty, and livelihood security are pivotal to Kerala's agricultural growth. However, to sustain these impacts, continuous enhancement of efficient of extension management practices/approaches is inevitable.

Implications / Strategic Recommendations

- Strengthen the use of digital platforms and social media for wider technology dissemination, particularly targeting younger audiences.
- Conduct of more HRD programs for scientists on emerging/niche areas
- Creation of more business incubation centres: Fostering innovation and entrepreneurship through the creation of additional business incubation centres, for providing a conducive environment for thc development of agri-businesses.
- Introduction of skill parks for providing higher order multi skill training to the end users for ensuring a more versatile and capable workforce
- Encourage/Promote consultancies aimed to enhance expert advice and guidance for farmers and stakeholders. It is proposed to bring out a consultancy policy of KAU
- Foster closer collaboration between the DoE, private sector, and community-based organizations to increase outreach and impact.
- Enhance/Strengthen feedback mechanisms from farmers and stakeholders to ensure that the initiatives of the Directorate of Extension remain responsive and adaptive to changing agricultural needs.

References

1. https://www.kau.in
2. https://www.extension.kau.in
3. https://www.cti.kau.in
4. https://www.celkau.in
5. https://ccmannuthy.kau.in

2

Digital *Choupal* of Bihar Agricultural University (BAU) for Knowledge and Frontline Extension Management

R.K. Sohane[1] and Priti Priyadarshni[2]

[1]*Director, Extension Education and* [2]*Assistant Professor-cum-Jr Scientist, Extension Education, BAU, Sabour*

Email: deebausabour@gmail.com

Abstract

With the rapidly increasing population, agriculture has kept pace with growing productivity demands. However, the widening gap between farmers and extension agents poses significant challenges for effective knowledge dissemination. In this complex scenario, ICT has emerged as a transformative solution. Its success, however, depends on affordability and the delivery of customized information tailored to the highly localized needs of agriculture, particularly during crises like the COVID-19 pandemic. To address these challenges, Bihar Agricultural University (BAU), Sabour, transitioned its Kisan Choupal model, launched in 2012, into the Digital Choupal. This adaptation enabled seamless knowledge exchange, real-time feedback, and personalized training for farmers. Over the years, this initiative has steadily gained the trust and attention of farmers.

Keywords: *ICT, Digital Choupal, Real time Farmer-Scientist Interface*

Introduction

The modernization of farming systems across the country is transforming agriculture. Yet, public extension services often struggle to keep pace with the growing demands for knowledge dissemination due to widen gap between farmers and extension agent *i.e.,* 5000:1 (Ragasa *et al.*,2013). Also, only 6.8% farmers receive extension services directly from state extension agency (GFRAS, 2012). The complexity of modern farming systems, encompassing decisions about land use, crop selection, and market access, underscores the importance of tailored specific knowledge. Strong public extension services typically reach only about 10% of farmers, a figure further constrained by limited operational resources (Bell, 2015). This gap is exacerbated by limited resources and infrastructure. In this context, Information and Communication Technology (ICT) emerges as a game-changer, offering innovative solutions to address these challenges and empower farmers with crucial insights. The integration of ICT in agriculture is revolutionizing the way knowledge and services reach farmers, especially in resource-constrained and remote regions. ICTs, such as mobile technology, have demonstrated significant potential in extending the reach of agricultural extension services. For instance, mobile hotlines now enable farmers to obtain technical advice and market information, including price updates and market locations (Aker and Mbiti, 2010). Such tools empower users, particularly in remote areas, to access vital information and communicate efficiently (Aker, 2011). To meet farmers' diverse and specialized information needs, ICT-based models must bridge these gaps effectively. However, the success of ICT interventions depends heavily on affordability and the customization of information to address the highly localized nature of agriculture (McNamara *et al.*, 2011; Bell, 2015).

Digital Choupal

Recognizing these challenges, Bihar Agricultural University (BAU), Sabour, has pioneered an innovative solution by introducing Digital *Choupal* through videoconferencing facilities at Krishi Vigyan Kendra (KVKs) and colleges of the university across the state. This initiative has supported agricultural extension by significantly reducing the barriers posed by distance and resource limitations. Bihar, a densely populated state in India with 1,102 people per sq. km, faces significant challenges in its agricultural sector. Employing around 75% of the workforce and contributing 35% to the GDP, agriculture remains a vital part of the state's economy. However, technology adoption is slow, particularly in rural areas, due to poor connectivity and limited communication infrastructure. Less than half of Bihar's villages are connected by roads, and access to trained extension personnel is scarce. These limitations hinder the dissemination of agricultural innovations and the provision of real-time advisory to farmers, which are essential for enhancing productivity and improving livelihoods. This innovative approach has redefined agricultural outreach, ensuring that even the most remote farmers can connect with experts and benefit from customized training and guidance.

Evolution of Digital Choupal

Recognizing the vital role of smallholder farmers in agriculture and the challenges faced by resource-poor rural families, Bihar Agricultural University (BAU), Sabour, launched the "Kisan *Choupal*" initiative on April 28, 2012 with following mandate.

Mandate of the Kisan Choupal

- To strengthen linkage between scientists and farming community and provide scientific information at farmers' doorsteps.
- To revive the tradition of Kisan *Choupal* existing in the ancient times to help farms solve their problems on their own at their place.
- To collect feedback and/or researchable issues from farmers' fields and communicate to the researchers for further formulating research priorities.
- To provide area specific/demand driven information of farmers at their doorsteps.
- To Bring about convergence between different agencies working for at grassroots level.
- To motivate the people with these of scientific and technical videos on farming practices.
- To create awareness about and popularize the ongoing government schemes for agricultural development

This effort was undertaken in collaboration with 20 Krishi Vigyan Kendras (KVKs) and 8 constituent colleges of the university. Since its inception, Kisan *Choupal* has been organized every Saturday under the theme "Bihar Krishi Vishwavidyalaya Kisano ke Dwar - Kisan *Choupal*" (Bihar Agricultural University at Farmers' Doorsteps). BAU became the first institution in India to introduce such an innovative and farmer-centric program.

The Birth of Digital Choupal

But COVID-19 pandemic disrupted in-person interactions, posing significant challenges for providing timely advisory services to farmers. To address this, BAU introduced the Digital *Choupal*- Real time Farmer-Scientist Interface. By leveraging digital platforms, this innovative solution ensured uninterrupted knowledge dissemination, timely advisory support, and fulfilment of farmers' requirements during the crisis. Digital *Choupal* not only overcame the constraints of the pandemic but also laid the foundation for a modern, technology-driven approach to farmer engagement, continuing to empower the agricultural community.

Figure 1 and 2 reveals Digital *Choupal* has enabled a more customized approach to training sessions. These sessions address the unique agricultural requirements of various regions and seasons, ensuring that farmers receive relevant and practical guidance. Topics covered range from Fisheries management, animal husbandry, weed management and poultry vaccination to orchard management, goat farming, nutrient management, and mushroom cultivation. Figure 3 reveals that fisheries farmers represent the largest group benefiting from this initiative (27%), followed by those in animal husbandry (20%) and horticulture (17%).

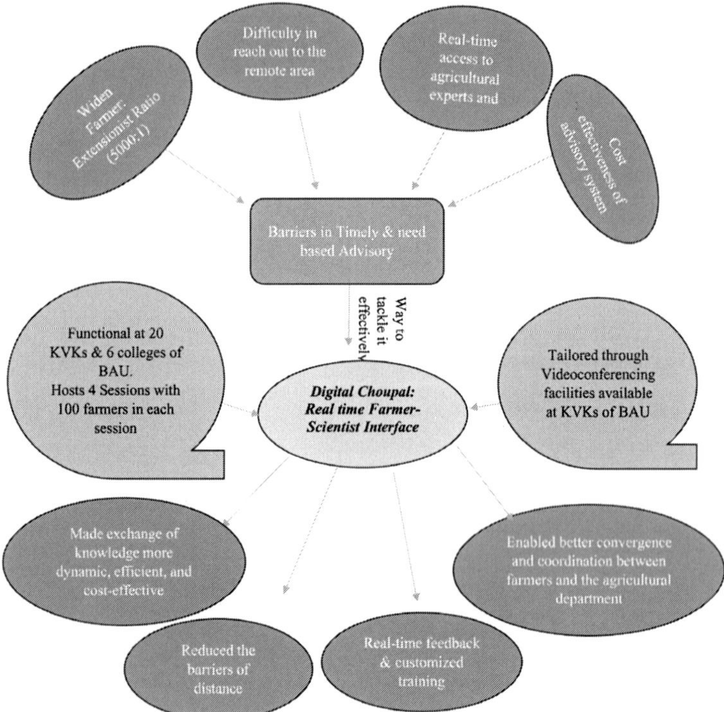

Fig 1. Conceptual framework of Digital Choupal Model of BAU, Sabour

Fig 2. Farmer-Scientist Interface

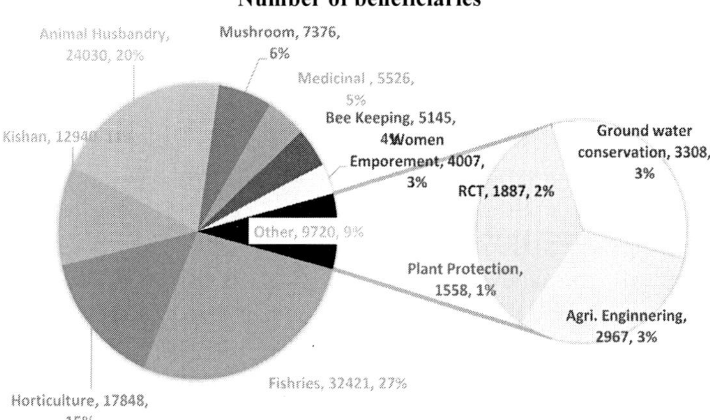

Fig 3. Number of participants from various disciplines

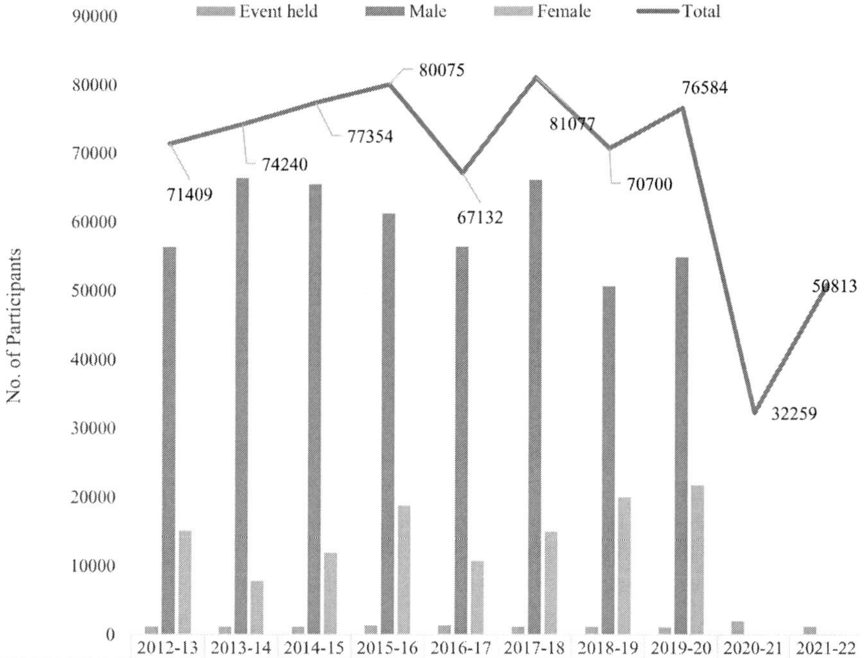

No. of Participants

	2012-13	2013-14	2014-15	2015-16	2016-17	2017-18	2018-19	2019-20	2020-21	2021-22
Event held	1176	1200	1200	1372	1392	1204	1199	1130	1976	1248
Male	56316	66419	65492	61241	56408	66119	50634	54832		
Female	15093	7821	11862	18834	10724	14958	20066	21752		
Total	71409	74240	77354	80075	67132	81077	70700	76584	32259	50813

Event Year

Fig 4. Number of participants and events

This tailored approach enhances the effectiveness of training and equips farmers with actionable knowledge. The program's reach is extensive, encompassing all 20 Krishi Vigyan Kendras and 6 colleges affiliated with the university. Each day, four videoconferencing-based training sessions are conducted, each accommodating approximately 100 farmers. Figure 4 illustrates the growing popularity of Kisan *Choupal* over the years. However, when it transitioned to Digital *Choupal*, the number of participants initially declined due to limited awareness, dropping from 76,584 in 2019-20 to 32,259 in 2020-21. Despite this, the initiative regained momentum and successfully benefited 50,813 farmers by June 2021 in the financial year 2021-22. This ICT platform also facilitated a significant increase in the number of events, rising from 1,130 in 2019-20 to 1,976 in 2020-21 (Figure 4). The sessions have remained impactful, incorporating practical demonstrations, outdoor broadcasting, and crop health diagnostics, which provide farmers with immersive, hands-on learning experiences. Furthermore, the program has fostered better coordination between farmers and the agricultural department. This synergy ensures that the knowledge imparted aligns with the specific needs and priorities of the farming community, enhancing the relevance and impact of extension services.

In a technological era where remote communication is crucial, the initiative has taken an additional step by incorporating live demonstrations through mobile signal aggregators. This initiative is gaining more momentum day by day by enabling farmers to observe agricultural practices in real time, even in areas with limited connectivity. It is enhancing the accessibility and effectiveness of the knowledge transfer process.

Conclusion

Digital *Choupal* has enhanced the agricultural outreach in remote areas. By facilitating seamless knowledge exchange, real-time feedback, and customized training, this ground-breaking approach empowers farmers with essential agricultural insights. It exemplifies how technology can bridge geographical gaps, transforming extension services and contributing significantly to the growth and sustainability of agriculture in Bihar. This model highlights the transformative potential of ICT in addressing the limitations of traditional agricultural extension systems. By tailoring information to farmers' needs and utilizing cutting-edge communication tools, the initiative serves as a beacon of innovation in agricultural development, ultimately improving the livelihoods of farming communities.

References

1. Aker, J.C. 2011. Dial "A" for Agriculture: A Review of Information and Communication Technologies for Agricultural Extension in Developing Countries. Agricultural Economics, 42: 631–647.
2. Aker, J.C. and Mbiti, I. 2010. Mobile Phones and Economic Development in Africa. Journal of Economic Perspectives, 24(3): 207-232.
3. Bell, M. 2015. Powering Behaviour Change for a Brighter Agricultural Future. MEAS Discussion Paper, University of California, Davis.
4. GFRAS. 2012. Fact Sheet on Extension Services. Position Paper. Global Forum for Rural Advisory Services (GFRAS) June 2012.
5. McNamara, K., Belden, C., Kelly, T., Pehu, E. and Donavan, K. 2011. Introduction: ICT in Agricultural Development.
6. Ragasa, C., Berhane, G., Tadesse, F., & Taffesse, A. S. 2013. Gender differences in access to extension services and agricultural productivity. *The Journal of Agricultural Education and Extension*, *19*(5), 437-468.

3

Innovative Extension Approaches Adopted by OUAT to Meet the Real Time Needs of the Farmers in Odisha

Prasannajit Mishra

Dean, Extension Education, Odisha University of Agriculture and Technology

Email: dee@ouat.ac.in

Abstract

The Directorate of Extension Education at Odisha University of Agriculture and Technology (OUAT) has implemented a range of innovative, inclusive, and technology-driven agricultural extension strategies to address the real-time needs of farmers across Odisha. This article explores the integrated approach adopted by OUAT through initiatives such as a structured Krishi Vigyan Kendra (KVK) action plan process, the establishment of a Centre of Excellence for Farmer Producer Organizations (FPOs), the conduct of Research-Extension Linkage Meetings to strengthen the feedback loop between scientists and farmers, and the deployment of the Krushi Samrudhi Helpline for real-time advisory services. These interventions collectively enhance the responsiveness, relevance, and reach of agricultural extension services in the state. The article further illustrates how these models ensure technology dissemination, capacity building, and farmer empowerment across diverse agro-climatic zones.

Keywords: *Agriculture, Directorate of Extension, Extesnion Services, Odisha*

Introduction

The Directorate of Extension Education, came into existence during 1967 under Odisha University of Agriculture and Technology. This directorate is engaged with various projects *viz*: University Extension Block Programme, Distance Education, Digital Extension, Information wing, Gender resource center, Educational Museum besides managing 31 KVKs in 28 districts of Odisha except Cuttack and Khordha district in which University Extension Block Programme and Farmers FIRST program is in operational. This directorate adopted several innovative extension approaches which made the

OUAT KVKs unique in operational modalities and effective in delivering for the development of farming community in Odisha, some of the innovative approaches are: Development of KVK Action Plan, CoE on FPOs, R-E Linkage and Krushi Samrudhi helpline.

Development Process of Action Plan

Krishi Vigyan Kendras work as a knowledge resource center in agriculture and allied field for the district (Singh *et al.,* 2012) where in, production and supply of critical technological products, both tangible and intangible is the major component other than having scientifically maintained demonstration units, crop cafeteria etc. Development of complete feasible action plan of KVKs requires sincere efforts in identifying the real problems on one hand and sourcing the frontline technologies on the other. The logical fitment of suitable technological options for testing, in addressing situation specific problems are also an important activity requiring our understanding of the major farming systems in the AES, micro-farming situations of different crops & commodities and our scientific orientation. For this OUAT adopts a uniform action planning process since last few years which involves different activities.

1. Identifying the problems
2. Prioritizing the problems
3. Analyzing the problems and finding root cause(s) of the prioritized problems.
4. Sourcing alternative technologies and listing possible solutions nearer to hit the root cause and address the problems.
5. Taking into consideration the follow up activities require to be undertaken for previous year OFTs and FLDs
6. Planning interventions including those for other extension activities, production of critical inputs etc.
7. Consolidation of interventions and finalization of action plan with respect to OFT, FLD, Capacity building, other extension activities or combination of these.

For all operational purposes, the process of Action Planning in KVKs are broadly divided into the following major steps:

Step–1: Scouting, consolidation, prioritization and analysis of the problem arriving at a problem matrix at each KVK level. (Through regular field visits, Farmers feedback, SAC suggestions etc.)

Step–2: Sourcing, consolidating and listing technology options arriving at a technology inventory for that district and more specifically for the

AES under operation. (Technology Facilitations Teams at University level headed by senior scientists / faculties from concerned field will support the process)

Step-3: Analyzing, evaluating and selecting suitable technology options to counteract the root cause and address the prioritized problem (Joint team of faculties and KVK Scientists)

Step-4: Listing interventions required as follow up activities for carrying forward last years' interventions to a logical end. (Discipline wise joint team of faculties and KVK Scientists)

Step–5: Planning for each individual intervention viz. OFTs, FLDs, Trainings, other Extension Activities including that for production of critical technology products. (Joint team of faculties and KVK Scientists)

Step-6: Consolidation and development of a logically linked and holistic action plan for the KVK. (Joint team of faculties and KVK Scientists)

Step-7: Finalization of draft action plan. Presentation before departmental expert team and State Level Extension Council.

Step-8: Finalization of Action Plan

Centre of Excellence for FPOs

The Centre of Excellence (CoE) for Farmer Producer Organizations (FPOs) is being operational at the Directorate of Extension Education is an initiative that aims to provide targeted, sustainable, and long-term support to FPOs across the state of Odisha. The CoE serves as a centralized hub for resources, knowledge dissemination, training, policy advocacy, and collaboration with key stakeholders, ensuring that FPOs achieve their full potential in the agricultural sector. The CoE is funded by the Government of Odisha, specifically through the Directorate of Horticulture, with the primary goal of enhancing the capacity of FPOs across the state.

The Centre of Excellence has defined objectives aimed at building a robust and sustainable FPO ecosystem in Odisha. The main objectives include unified FPO ecosystem, capacity building, market and credit linkages, knowledge sharing and digital tools and collaborative decision-making for smart procurement of inputs as well as smart marketing of produce.

The CoE for FPOs operates in close collaboration with several stakeholders, including the Government of Odisha, OUAT, KVKs (Krishi Vigyan Kendras), Implementing agencies, CBBOs (Cluster Based Business Organizations), credit institutions, market experts, and various other developmental organizations. Its mode of operation includes training programs, workshops, exposure visits, mentoring and coaching, technical support and digital resources.

The Project Management Unit (PMU), based at the Directorate of Extension Education (OUAT, Bhubaneswar), plays a pivotal role in the coordination and management of the Centre of Excellence. The PMU is responsible for coordination, monitoring and evaluation, financial management, reporting, training and capacity building. The PMU also ensures that regular reviews and evaluations are conducted to assess the effectiveness of the project's activities and its impact on the FPO ecosystem in the state.

The KVKs play a key role in the grassroots implementation of the CoE's activities. Each KVK is tasked with the responsibilities like FPO selection and nurturing, training programs, on-field demonstrations, exposure visits, technical support and documentation and reporting.

The Centre of Excellence for FPOs at OUAT plays a pivotal role in empowering FPOs across Odisha, strengthening their capacity, fostering collaboration, and ensuring their sustainability. By leveraging the resources of the government, OUAT, KVKs, and other stakeholders, the CoE provides a comprehensive, multi-dimensional approach to address the challenges faced by FPOs. The ongoing efforts and activities of the Centre will play a vital role in shaping the future of agriculture in Odisha, making FPOs more productive, financially stable, and socially responsible.

Research-Extension (RE) Linkage Meeting: An Innovative Extension Method for Technology Transfer in the State

The concept of technology flow is fundamental to the design of research and extension systems. This facilitates diagnosis of research-extension linkage problems. Research and extension have their own priorities, which determine their main activities. The issues like non-adoption of recommended technologies arise because of weaknesses in the links between research and extension institutions. Technology development and transfer functions are treated in isolation. Reasons for the need for strong linkages include influencing formulation of research agendas based on problem identification, and the need to evolve technology suitable for the prevailing socio-economic and ecological environment. Extension can provide information and facilitate interaction between researchers and farmers. Simultaneously, extension also requires a constant flow of information on new and improved practices, necessitating a two-way communication process. In a research organization, technology integration and production activities are too often neglected. For example, specialists of different subjects try to develop technologies related to their respective fields and do not integrate the same to come up with a modular form. Usually, a scientist goes farthest in testing the technology generated and often with a pro-invention bias that his/her technology will be universally adopted. In contrast, the extension activities are swivelling around the available technologies and many a times do not find a suitable technology for a particular intervening point. Therefore, the most critical linkages are at the stage of technology integration and again at technology testing (assessment/refinement) and dissemination. The linkage problems have not been appreciated until recently. However, there is a need to strengthen the research-extension linkage in circumstances of diverse ecological conditions because the linkage is often good in high yield/production systems. It defies the very own hypothesis of complete adoption of technologies developed by any research institute which will eventually tackle the problems like research innovation paradox.

Rationale of the meeting are (i) Research problems being investigated are generally not in accordance with the priority needs of agricultural producers, and (ii) Knowledge generated at research stations has not been effectively transferred to the producers. The objectives are (i) Transfer of research results, innovative technology/information suited for the concerned farming systems of the state in general and district in particular, and (ii) Taking real time feedbacks from all the stakeholders including the farmers/farmwomen present in the meeting.

The meeting is being held in yearly basis at the Department of Agriculture and Farmers' Empowerment or University for State-level and monthly on a fixed day in the KVK for District Level. Here each participating body represents one principle and the entire members act as a system. The State level R:E meeting is chaired by the Hon'ble Vice Chancellor of OUAT. This is also named as Apex committee in which Secretaries and Directors of all related line departments Dean of Research and Dean Extension are members.

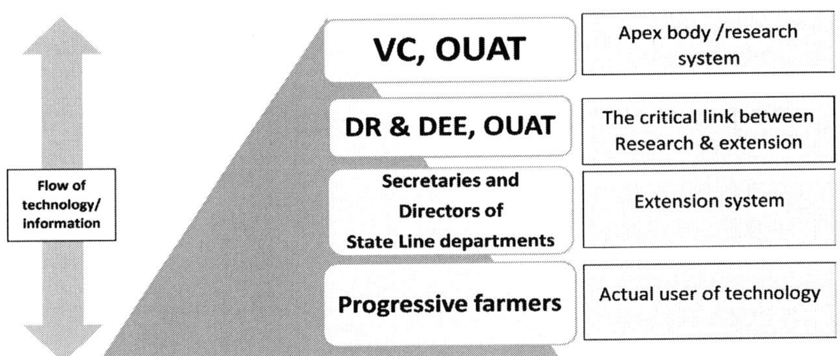

Fig 1. Visual representation of R-E interface meeting at state-level

The District Level R: E meeting is chaired by Associate Director of Research of that Zone. The senior scientist & head of KVK is the convener of the meeting. The burning issues of the district/farmers were discussed in the monthly meeting. Then the appropriate technologies are provided, if available. If not, the problem directly goes to the research system. Farmers' feedback is discussed here for the available technology or new problems. The presences of line department officials make the horizontal expansion easier. The inclusion of all parties in the method makes it the 'increasingly integrated research extension producer interaction system'. The ultimate aim of this method is to address the issues like innovativeness-needs paradox thus reducing the growing socio-economic gap in long run.

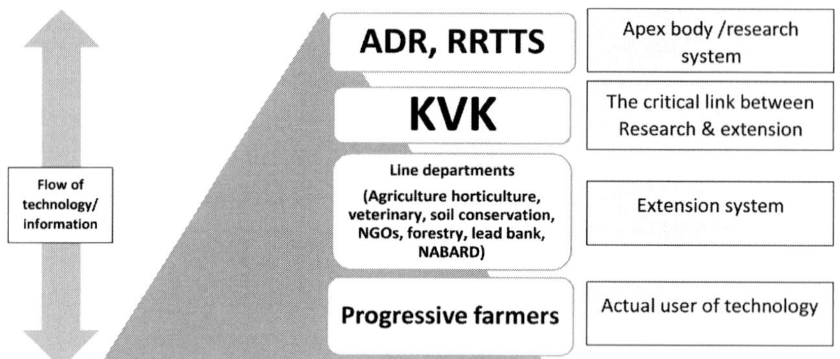

Fig 2. Visual representation of R-E interface meeting at district level

Krushi Samrudhi Helpline

Krushi Samrudhi Helpline is a free two-way mobile phone-based advisory service offered by the Department of Agriculture and Farmer's Empowerment and Fisheries & Animal Resources Development Department, Government of Odisha with technical guidance and support from Directorate of Extension Education, OUAT.

Objective

This program is designed to cater to the farmer's queries & timely dissemination of customized information on farming practices. Under this program, farmers are provided with customized advisory on Agriculture, Livestock and Fisheries Management. This program started in July 2017, and subsequently through transition transferred to Tatwa Technologies Ltd. in April, 2022.

The Krushi Samrudhi Helpline Call Centre

There is a 100-women-empowered Call Centre that receives calls from farmers and the public from all over the state. These women are well-trained by the department to address the queries of the farmer & the public to provide a satisfactory response. Periodic trainings are conducted by the department to update the staff on revised updates on schemes and other programs. Tailor-made advisories are disseminated based on a scientific and improved crop (Agriculture) and events (Livestock & Fisheries) calendar vetted by Content Review Committee. Which additionally includes Government notifications, various farm practices, weather alerts, and scheme related information on a timely basis. A repository of contents is created for future reference. Live telephonic interaction with experts introduced for greater satisfaction of Farmers as recommended by DAFE & FARD. Krushi Samrudhi Helpline program facilitates farmer with Live Call Centre where the farmer can dial a toll-free number-155333 and get information on various agricultural practices. 12 Agriculture experts, 5 Livestock experts and 3 Fisheries experts answer to the technical question of the farmers. Farmers also register their grievances related various subjects at the Live Call Center. The Live Call Center answers around 50K calls monthly.

Activities

Inbound Service: Farmers ask their questions on farming to the live call center by dialing the toll-free number 155333. The Call Center agents receive and provide a response to the farmer on the same call. All the technical questions on farming are transferred to the experts for a one-on-one telephonic discussion to provide a firm resolution.

Domain Expert's one-on-one Consultation with the Farmers

Upon transfer of farmer's calls from the KSH agents, the respective Domain experts engage in an intense discussion with the farmers to understand the farmer's issue in farming and provide appropriate resolution vetted by the department.

Outbound Service

- A call is made to the farmers for profiling to add them to the database.
- Impact evaluation of Government schemes implementation by feedback surveys.
- Scheme beneficiaries nudge and sensitized through individual farmer calling.
- Extension worker compliance nudge to meet deadlines

Interactive Voice Response System: Seasonal and timely voice call advisories are broadcasted to the farmers throughout the farming season to help farmers make informed decisions on their farming. Emergency weather alerts are disseminated to the farmers about incessant rains, high temperatures, cyclones and storms so that they protect and secure their crops. Likewise, fishermen are warned not to venture into the sea and rivers during cyclones and storms. IVRS service is additionally provided to APICOL, RFM, REWARD, and Soil Conservation for dissemination of customized advisories.

Advisory Content Creation: The content of the advisory is created by the domain experts of Agriculture, Livestock and Fisheries with reference to sources recommended by the department.

Advisory Vetting: The advisories are vetted by members of the Content Review Committee under the Chairmanship of Dean Extension, OUAT Bhubaneswar

Advisory Recording: The vetted advisory is given a voice by the domain experts by recording the advisory in a clear, prominent voice.

Advisory Broadcast: Vetted advisory is broadcasted as per the schedule of the crop and events calendar.

Crop and Event Calendar: A scientific Crop & Event calendar is designed based on the timing of the farming activities of Agriculture, Livestock & Fisheries. The department vets this calendar to disseminate messages as per the time mentioned for any activity.

Currently, the Krushi Samrudhi Helpline disseminates advisories on the following:

- 66 different types of Agricultural Crops
- 11 types of Livestock
- 03 types of Fishes

On-call Artificial Insemination: Artificial Insemination in cattle, a new initiative taken by the ADR - Animal Resources Department is facilitated by KSH, currently operating in 10 districts allows farmers to directly book calls through their mobile phones at the ease of their home by calling KSH Call Center at 155333 which helps them for quick and faster execution of insemination for their cattle.

SMS Blast: For rapid dissemination of messages to the farmers upon directions from the department SMS blast is triggered. A large mass of farmers can receive important, short-noticed information through SMS blasts.

Field Team: KSH has a Field team led by a Field manager to sensitize farmers and train the extension workers to help farmers use the KSH service on the ground in all 30 districts. The Field team popularizes KSH among the farmers encouraging them to use its services to get the optimum benefit in their farming. The extension workers are trained to use AI Chatbot to seek resolutions for farmers approaching them with queries on their farming. Till date in total approximately 1.30 billion advisories are disseminated to Farmers.

Table 1: Details on advisories disseminated

Advisory Created	**Till Mar 22**	**Apr 22-Nov'24**	**Total**
Agriculture	246	1150	1396
Livestock	190	400	590
Fisheries	90	350	440
Total	526	1900	2426
Krushi Samrudhi Farmers Count	32 Lakh	42.72 Lakh	74.72 Lakh
Agriculture	20	46	66
Live Stock and Fisheries	8	6	14

Data Team: A dynamic data team works in the backend to manage the software and applications involved in running the KSH Call Center. Its efficient data management system enables it to pull out every detail of the advisory or information disseminated through IVRS or SMS blasts.

Tech Team: A technically sound tech team works hard to maintain the day-to-day glitches in the hardware & networking functions of the systems that enable KSH to operate, store & secure data digitally. It ensures stable internet access to the systems for smooth operations.

Working Hours: The KSH operations provide a 12-hour service window from 8 am to 8 pm 365 days a year to the farmers on its toll-free number 155333.

Reference

1. Singh, K.M., Singh, P., Shahi, B. and Shekhar, D. 2012. Role of Krishi Vigyan Kendras (KVKs) in Agricultural Extension: An Overview.

4

Agri-Informatics and Agri-Electronics for Socio-Economic Development of Farmers

B.K.Jha[1], Neetu Kumari[2], Niva Bara[3], J.Oraon[4] & Ritesh Mukherjee[5]

[1]Head, Department of Agricultural Extension Education, [2]Assistant Professor-cum-Junior Scientist, Department of Agricultural Extension Education, [3]Former Head, Department of Agricultural Extension Education, [4]Director of Extension Education, Birsa Agricultural University, Ranchi, [5]Associate Director, Centre for Development of Advanced Computing, Kolkata

Email: basantbkjha@gmail.com

Abstract

Information and Communication Technologies (ICTs) have transformed the society, economy and polity. Both Government and Non-Government Organizations have been taking initiatives to keep pace with the ever-evolving ICTs. Birsa Agricultural University, Ranchi have taken initiatives from time-to-time to integrate ICTs in teaching, research and extension education. It has developed multilingual technology portal on agriculture, livestock and forestry; an artificial intelligence based multimodal dialog system for farmers; established community radio station (CRS) and launched YouTube Channel (BAU EXTENSION). In next phase of its pursuit, it established Technology Resource Centre (TRC) on Agri-informatics and Agri-electronics. This centre comprises achievements of earlier initiatives and laboratory for agri-electronics equipments for assessing quality of agricultural produce. Over 45,000 people have visited technology portal and piggery has emerged as the most important technical content on YouTube channel with viewership of about 16,000. Being ICT based initiative; the reach of the system is worldwide. Farmers of the nearby areas are showing interest in community radio and participating in the program in a big way. Stakeholders are visiting TRC laboratory and a few processors have shown interest in installing the agri-electronics equipments in their units. Motivated by the response of farmers, FPOs, NGOs, Entrepreneurs, officials and industrialists, the university will strive to go for the integration of Internet Of Things (IOTs), block chain and big data analytics.

Keywords: *Technology portal, Artificial Intelligence, Agri-informatics, Agri-electronics, Community Radio, YouTube*

Introduction

Information and Communication Technology (ICT) has emerged as a transformative force in agriculture, significantly enhancing access to information, improving productivity, and promoting sustainable practices. The integration of ICT tools facilitates the dissemination of critical agricultural knowledge, enabling farmers to make informed decisions regarding planting, pest control, and market trends. This technological advancement is particularly beneficial for small-scale farmers who often face challenges in accessing timely and relevant information. ICT platforms such as mobile phones and radios play a crucial role in providing small-scale farmers with access to essential agricultural input information, aiding them in making informed decisions (Fosu & Giba-Fosu, 2024). Furthermore, ICT tools, including e-learning platforms and mobile applications, have revolutionized agricultural extension services by providing direct access to advisory services and expert knowledge.

The use of social media and mobile communication has also improved interactions between farmers and agricultural experts, thereby increasing the effectiveness of extension programs (Bhat *et al.*, 2024). The advent of digital and smart agriculture, supported by ICT, incorporates technologies such as remote sensing, precision agriculture, and automated machinery to optimize farming practices and reduce labor dependency (Yoe, 2024). Agri-informatics, a specialized branch of ICT in agriculture, leverages emerging technologies like Big Data and the Internet of Things (IoT) to develop sophisticated agricultural systems, promoting smart farming practices (Jena, 2024). While ICT provides numerous advantages, challenges such as infrastructural limitations, economic constraints, and disparities in digital literacy hinder its widespread adoption. Addressing these barriers through policy interventions and capacity-building initiatives is crucial for maximizing the potential of ICT in agriculture.

Artificial Intelligence (AI) is increasingly integrated into agriculture to address contemporary challenges such as climate change, population growth, and the demand for sustainable food production. AI applications in soil and crop monitoring, predictive analytics, and agricultural robotics are gaining attraction, assisting farmers in data collection and analysis (Smith, 2020; Gupta *et al.*, 2023). Moreover, automation in agriculture, driven by AI, addresses issues such as crop diseases, irrigation management, and pest control. Techniques like IoT, machine learning, and deep learning contribute to the automation of farming practices, increasing soil fertility and productivity (Jha *et al.*, 2019). Robotics and automated systems minimize human intervention, thereby enhancing farm efficiency and sustainability (Eli-Chukwu, 2019). The implementation of AI in agriculture yields numerous benefits, including enhanced productivity, cost-effectiveness, and sustainability. AI assists in optimizing resource utilization, such as water and fertilizers, and contributes to the reduction of greenhouse gas emissions (Shaikh *et al.*, 2022). However, the adoption of AI in agriculture faces challenges such as data replicability, privacy, and security concerns.

The inherent variability in field conditions and farming techniques complicates systematic data gathering and analysis (Sahoo & Sharma, 2023). Additionally, the introduction of AI has the potential to disrupt traditional roles and skills in farming, necessitating careful consideration of its social and ethical implications (Svetskiy, 2022). Future research in AI-driven agriculture will likely focus on developing autonomous and intelligent robots for tasks such as plant and soil sample retrieval and effective livestock management. The integration of AI with digital twins and supply chain data is expected to further enhance decision-making and farm management (Patel & Patil, 2022). Continued advancements in AI technologies, particularly deep learning, will improve the processing of large datasets, enabling more intelligent and timely agricultural decisions (Belattar *et al.*, 2023).

Birsa Agricultural University, Ranchi has developed multilingual technology portal; launched YouTube channel "BAU EXTENSION" and established community radio station for the benefit pf the farmers, entrepreneurs and extension functionaries. Keeping pace with changing technology, the university applied artificial intelligence (AI) in information access by the farmers. The efforts culminated into establishment of Technology Resource Centre on Agri-informatics and Agri-electronics in collaboration with Centre for Development of Advanced Computing, Kolkata.

About the Technologies

i. **Context and Background:** The dimension of extension has widened to include ICT, secondary agriculture, value addition, agri-quality, market and environmental issues. Agri-informatics has been changing the face of agricultural extension. Likewise, agri-electronics is slowly occupying space in overall farm operations. To shape the future of agriculture and making the system up-to-date, transparent and vibrant, the infusion of agri-informatics and agri-electronics is the need of the hour.

ii. **Problems Addressed:** Scientific agricultural information is not only influencing production but also the profitability of the farmers. At the same time quality of agricultural produce has been critical in deciding price of the commodity. Through agri-informatics infrastructure, farmers have been enabled to access information in text, audio and video format through internet and Mobile app. The facility created for assessing the quality of the agri produce will enable the farmers to fetch premium price.

iii. **Aims and Objectives:** Basic objective of the initiatives is to develop agri-informatics and electronics infrastructure at BAU, Ranchi which will be used for training, demonstration and service with the aim of shaping future agriculture. The specific objectives are:

- Information access through multilingual technology portal, radio, Mobile App and social media
- Assessing quality of agricultural produce through agri-electronics infrastructure like Annadarpan, E Tongue, Biosensing System for pesticide, E Nose for rice, CT- view for chilli and turmeric, e Quality Veg, and Hand-held E-nose.
- Training of stakeholders like Government officials, extension functionaries and progressive farmers including farmers' organizations
- Information and diagnostic services to the farmers and agripreneurs.

iv. **Strategy:** The proposed centre is unique in its kind which will endeavour to empower the farmers to solve many of the production and marketing problems. The details are hereunder:

a) Agri-informatics

1. Multilingual Technology Portal

Birsa Agricultural University has developed multilingual technology portal in collaboration with Centre for Development of Advanced Computing (CDAC), Kolkata.The portal aimed at attaining sufficiency in information access by the farmers and extension functionaries in respect of adequacy of content, languages of the people and mode of dissemination.

Languages: English, Hindi, Nagpuri, Santhali, Mundari, Kurukh and Ho.

Mode of dissemination: Internet, Mobile, Radio and off line LCMS CDS.

Content: Agriculture, Livestock, Forestry.

In the backdrop of feedback from farmers and extension functionaries, features have been added in the portal which are given below:

Special Features of Agriculture Portal

i. Crop - General and variety description

ii. Soil health and field preparation - Fertility map-based fertilizer recommendation and different types of field preparation

iii. Fertilizer - Recommeded fertilizer dose of different crops along with fertilizer calculator

iv. Water management - Water requirement and schedule of irrigation

v. Crop protection - Identification of pest and disease by picture and remedial measures
vi. Agricultural implements - Use and efficiency of implements
vii. Post-harvest - post-harvest and value addition of crops
viii. Market - List of input dealers
ix. Finance - Agricultural loan schemes
x. FAQ

Special Features of Livestock Portal

i. Livestock - General and breed description of livestock
ii. Reproductive management - Issues related to reproduction
iii. Nutrition - Feed requirement of animal and birds
iv. Housing - Different types of scientific housing
v. General Management - Care and maintenance of livestock
vi. Health Care - Symptoms of disease and preventive health care

Special Feature of Forestry Portal

i. Plantation - Productive and protective plantation
ii. Agro-forestry - Different types of agro-forestry systems
iii. Wasteland Management - Different types of wastelands and their management
iv. Watershed management - Catchment and command areas along with their management
v. Medicinal plant - Details of medicinal plant along with cultivation practices
vi. Non-timber forest product - Different types of NTFP
vii. Natural Forest - Status of natural forest

Table 1. Multilingual technology portals and their URLs

S. No.	Particular	URLs
1.	Multilingual Agriculture portal	http://birsakrishiseva.in/submit/
2.	Multilingual Livestock portal	birsakrishiseva.in/faces/weaai.jsp?form1:hyperlink2_zsubmittedLink=form1:hyperlink2
3.	Multilingual Forestry portal	birsakrishiseva.in/faces/weaai.jsp?form1:hyperlink3_submittedLink=form1:hyperlink3

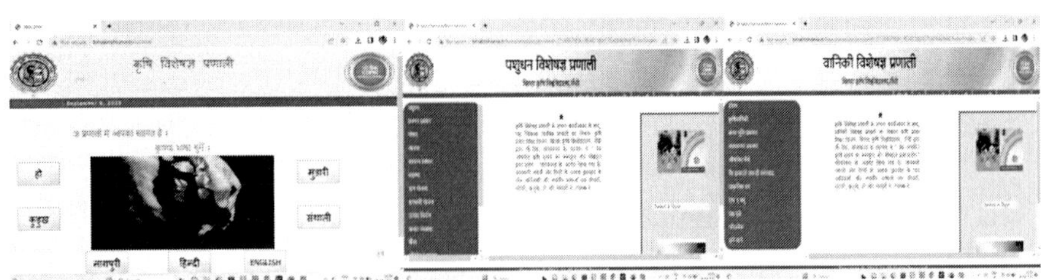

Fig 1. A view of multilingual technology portal

2. Application of Artificial Intelligence (AI) in Information Access

Birsa Agricultural University, Ranchi in collaboration with Centre for Development of Advanced Computing (Noida and Kolkata) and Bihar Animal Science University, Patna implemented a project "Krishi Mantrana: An AI based Multimodal Dialog System for the Farmers".

Features

i. Automatic Speech recognition (ASR): The application recognizes voice of speaker
ii. Speech to text (STT): The recognized voice is converted into text
iii. Text to speech (TTS): The identified text is converted into speech

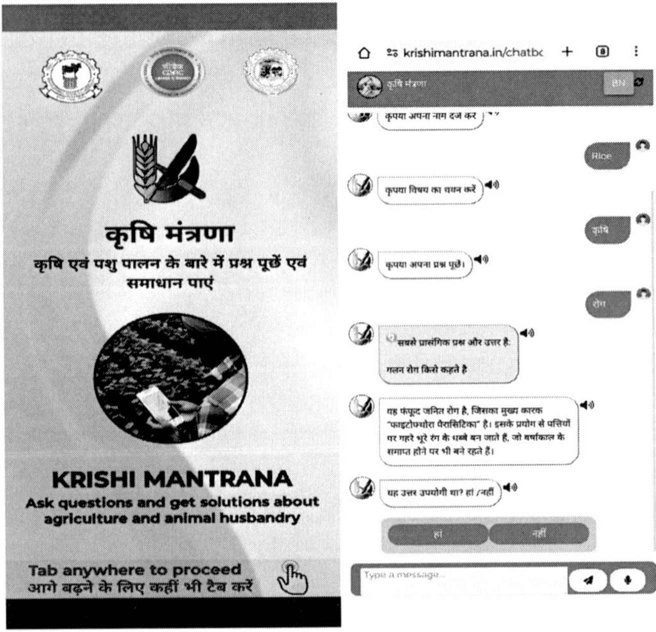

Fig 2. A view of Krishi Mantrana Mobile App

3. Use of Social Media in Extension

Social media facilitates interactions between people, enabling them to create, share, and exchange knowledge and ideas within virtual communities and networks. Increasingly, social media has become a powerful medium for sharing information and creating awareness. Platforms such as Facebook, What's App, YouTube, X, Instagram and telegram etc. have been used to engage audience globally. Worldwide 5.16 billion people are using social media which is 59.30 per cent of world population whereas in India 862 million people are active users constituting 59.90 per cent population. Facebook is the most widely used social media platform with user base of over 3 billion followed by YouTube (2.5 billion) and Instagram and WhatsApp with user base of 2 billion each. Birsa Agricultural University has well developed YouTube channel "BAU EXTENSION" with multilingual contents. The channel has over 150 films majorly in local language of the state.

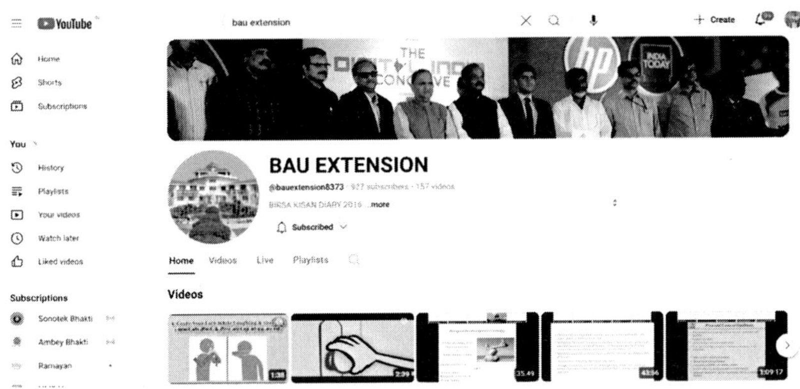

Fig 3. A view of home page of BAU EXTENSION

4. Community Radio in the Service of Farming Community

Community Radio has emerged as one of the important communication channels that can be used effectively to bridge the information gap existing in the farming community. Birsa Agricultural University, Ranchi has established a community radio station (CRS) in the name of "Birsa Hariyali Radio". It operates at 107.8 FM and broadcasts programs for three hours between 11.00 AM to 2.00 PM.

a) Agri-electronics

The university has established quality assaying lab comprising agri-electronics in collaboration with Centre for Development of Advanced Computing (CDAC), Kolkata. The details of equipment's along with their special feature are hereunder:

1. Annadarpan System for Rice and Pulses

ANNADARPAN ^{SMART}is a computer operated, portable device for appearance-based quality analysis of multiple crops like Rice, Tur dal, Bengal gram, Moong etc. Annadarpan^{SMART} system consists of a high resolution commercially available overhead scanner with built in high intensity LED based illumination, placed inside a detachable cabinet and a large sample handling tray with manual spreading of sample. Overhead scanner is interfaced and triggered from the software for image acquisition. Image analysis techniques have been implemented to analyze the sample as per the quality standard defined by Bureau of Indian Standards (BIS). Crop specific, separate software has been developed for appearance-based quality analysis.

Fig 4. Annadarpan^{SMART} system

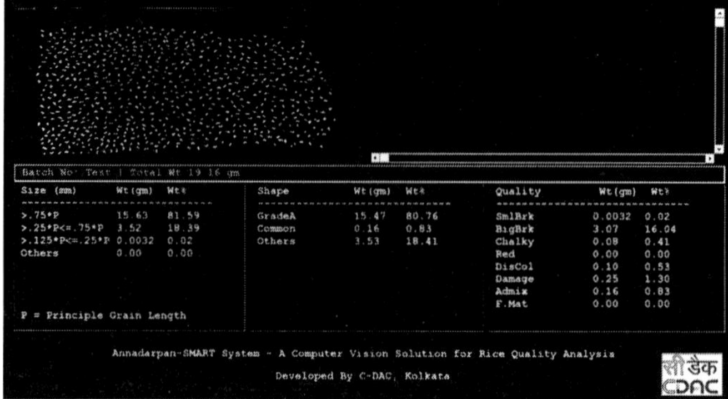

Fig 5. Software main screen of Annadarpan^{SMART}

2. Electronic Tongue

The E-Tongue system is based on voltammetry technique. Voltammetry applies a constant and/ or varying potential at an electrodes surface and measure the resulting current with a three-electrode system. These electrodes are working electrode, counter electrode and reference electrode.

i. Working Electrode: The working electrode is the electrode in an electrochemical system on which the reaction of interest is occurring. The working electrode is often used in conjunction with an auxiliary (counter) electrode, and a reference electrode in a three-electrode system. Depending on whether the reaction on the electrode is a reduction or an oxidation, the working electrode can be referred to as either cathodic or anodic. Common working electrodes can consist of inert metals such as gold, silver or platinum, to inert carbon such as glassy carbon or pyrolytic carbon, and mercury drop and film electrodes.

ii. The Auxiliary electrode, often also called the counter electrode, is an electrode used in a three-electrode electrochemical cell for voltammetric analysis or other reactions in which an electrical current is expected to flow. The auxiliary electrode is distinct from the reference electrode, which establishes the electrical potential against which other potentials may be measured, and the working electrode, at which the cell reaction takes place.

iii. A Reference electrode is an electrode that has a stable and well-known electrode potential. The high stability of the electrode potential is usually reached by employing a redox system with constant (buffered or saturated) concentrations of each participant of the redox reaction.

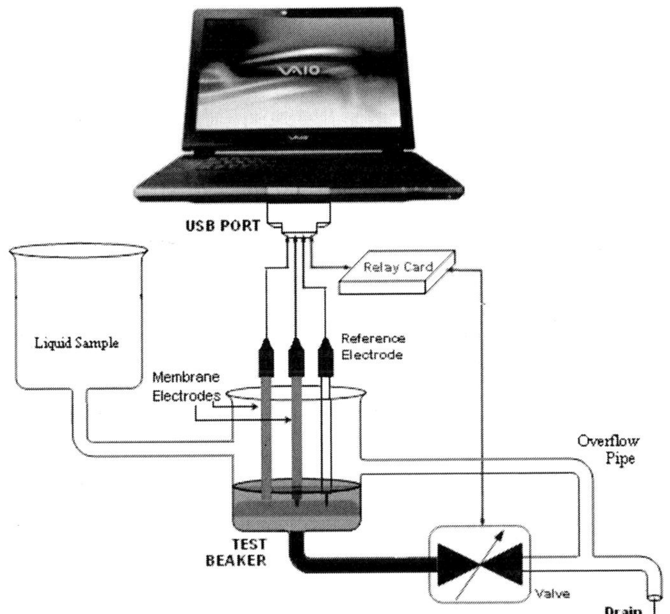

Fig 6. Block diagram of system

3. Biosensing System for Pesticide Detection

Agricultural pest management involves the use of a number of pesticides such as organophosphates (OP), organochlorines, synthetic pyrethroides, neonicotinoides etc. The most prominent of the OP pesticides which are abundantly used in the fields in the unorganized sector are chloropyriphos, ethion, monocrotophos, dimethoate, quinalphos, profenophos and phosalone in general.

Being the banned and most abundantly used pesticide, Monocrotophos (MCP) has been primarily focused for the present study. The detailed works done; activity wise is mentioned below:

Table 2. Features of merit

Features	$PesT^{SCAN}$	Conventional method
Range of detection	0.1 µg L^{-1} – 20 µg L^{-1}	-
Analysis time	30 min – 2 h	~5 days to 1 month
IC_{50}	1 µg L^{-1}	-
R^2	0.9966	-
Minimum detection limit	0.01 µg L^{-1}	2-5 µg L^{-1}
Total sample volume	100 – 300 µL	500 mL
Cost per sample (in INR)	400.00/- (Initial) 100.00/- (Bulk Manufacturing)	~5,000.00/- or 9,800.00/-

Device Interfacing

Uniform Illumination based Imaging System (UIIS) was interfaced with the PC via USB cable. The UIIS captures the images with 200 dpi, which is sent to PC via USB. When user presses the "Load" button, it grabs the image and converts it to 24-bit bmp ® image format for storing and further analysis.

Software Operation

a) This module is responsible for taking user input, selecting the mode of application (i.e. online or offline), selecting the reference sample/ testing sample.

b) The software has 96 number of colour palette from which the user can select the reference sample position and testing sample position corresponding to sample tray.

c) If user selects the reference at the time of testing, then the pesticide value will be given based on the reference sample data.

d) If reference is not available, user may also continue testing without using any reference.

e) It also gives a PC Index with the pesticide score.

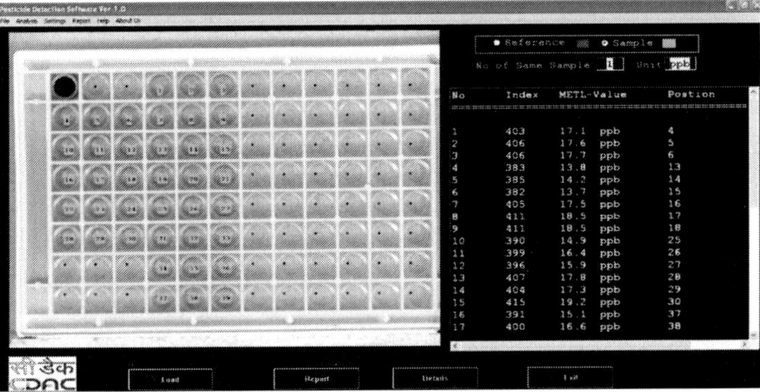

Fig 7. Software screenshot

4. Electronic Nose for Aromatic Rice

The quality analysis of rice involves different measurement parameters; AMMAR has been developed for assessing aroma. Through the advanced software of E-Nose system, AMMAR can be used for detecting and quantifying the aroma intensity in a digital form of aromatic rice, which is a very crucial parameter in rice quality domain from commercial angle. AMMAR - Electronic Nose system is capable of to detect the complex aroma of rice sample through an array of gas sensors and pattern recognition model. The quality analysis of rice involves different measurement parameters like dimensional analysis, appearance, aroma and chemical analysis. E-Nose and its compatible software have been developed for assessing aroma. Detection of aroma in a digital form is a very crucial parameter in rice quality domain, because the aroma of aromatic rice is the key factor from commercial angle.

Fig 8. Software interface arrays

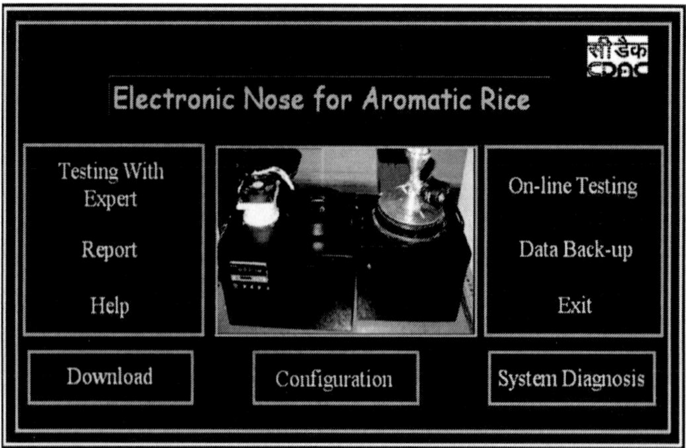

Fig 9. Software main screen

Electronic Nose for Aromatic Rice is capable of to detect the complex aroma of Rice sample. Where rice is prepared in Sample Preparation Unit & then the cooling & aroma sensing is done in Sensing Unit. Sample preparation procedures is given below:

a) Weigh 15gm. of Rice sample.
b) Pour it into the Sample Container with 60 ml. of distilled water.
c) Lock the sample container with sample holder & put it to Sample Preparation Unit (Unit 1).
d) Let's wait for 30mins to cook the rice.
e) Now shift the Sample Holder (with sample container) to Sensing Unit (Unit 2).
f) Let's wait for 15mins to cool down the sample in cold water.
g) Now mount the sensor array to the top of Sample Holder.
h) Let's open the software for testing & data collection.

The world's first Digital Rice Aroma Index Software not only gives score of the aroma, but also it can discriminate the non-aromatic rice with aromatic ones. This indexing method may help to study the chemical properties of rice aroma.

5. CT-VIEU for Chilli and Turmeric

CT-VIEU System (A Conveyorized Vision Inspection System for Quality Analysis of Dry Chilli Samples) is a conveyorized system with built-in mechanical feeding arrangement from a vibrating metal hopper wherein samples of dry chilli are fed and output of the feeding system is "delivery of sample dry chillies" on the top surface of a blue colour flat conveyor belt. Side metal frame of the conveyor is used to accommodate placement of Imaging Unit on top of the conveyor belt with necessary fixing arrangements on two side flanges of the frame. Imaging unit consists of a digital camera with illumination arrangement fitted inside an enclosed cabinet. Digital camera is connected with a computer through a standard universal serial bus (USB) interface and use to capture the live image of the chilli sample passing on the conveyor belt. Image analysis software running on a computer is used to inspect the quality of chillies, i.e. pod length greater than 5 cm, pod length between 3 to 5 cm, broken chillies, discolour, pod without stalk, loose seeds, foreign matters (stalks) and represents in terms of percentage weight. Dry chillies under test are collected through the output chute in a sample collector. The system is used for quality assaying of chilli sample for the purpose of quality inspection in the quality control (QC) laboratories. Measurable Quality Parameters are specified below.

a) Pod length (greater than 5cm)
b) Pod length between 3 to 5 cm
c) Broken chilli (based on size)
d) Pod without stalk
e) Discolour pod
f) Loose seed
g) Foreign matter (stalk)
h) Good quality

Fig 10. CT-VIEU system

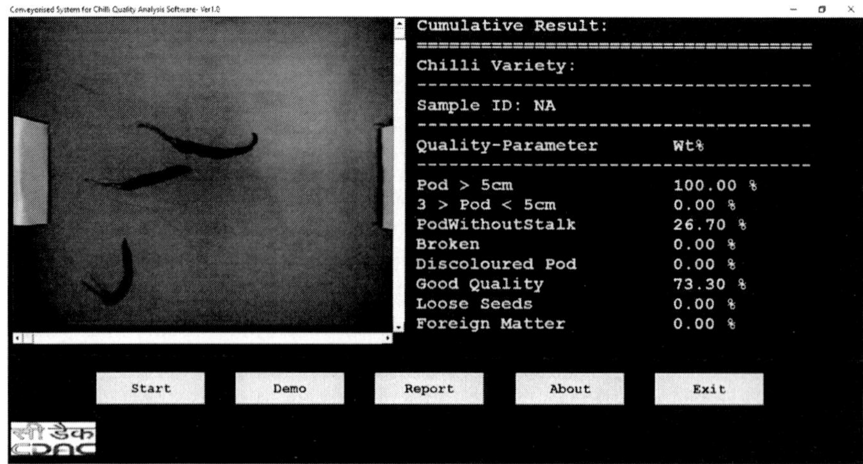

Fig 11. Software main screen of CT-VIEU

6. Conveyorized Vegetable Assaying Machine

eQuality^{VEG}is conveyorized system with a low-cost web camera and a low powered LED array based uniform illumination setup developed for analysis of vegetable like Potato, Tomato etc. Vegetable sample drops from the hopper on the conveyor belt such that one sample will be analyzed at a time. A method has been developed to avoid capturing the same frame twice by the camera. Software has been developed to calculate the quality parameters of vegetables as per AGMARK standard like Extra Class, Class I, Class II, Class III etc based on size. Crop specific, separate software has been developed for appearance-based quality analysis.

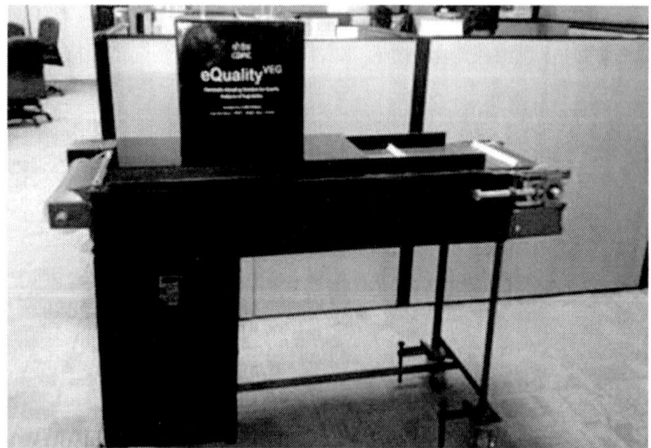

Fig 12. Prototype of e Quality^{VEG}

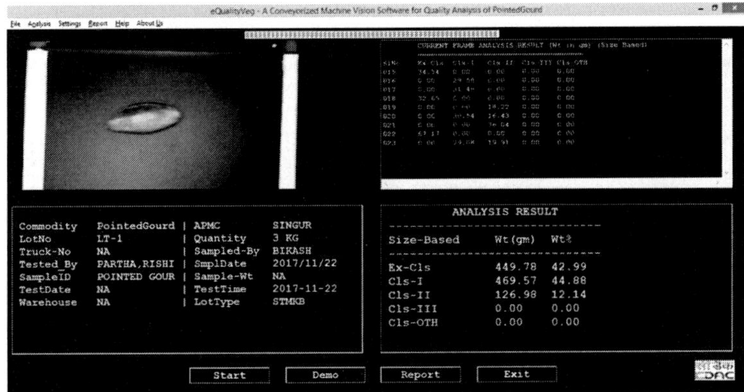

Fig 13. Software main screen of e Quality^{VEG}

7. Hand-held E-Nose

Handheld Electronic Nose (HEN) is an intelligent sensing and analysis device that can assess the quality of finished tea based on its aroma. HEN can also be used to determine the optimum fermentation point during tea manufacturing process. The sensors of HEN are an array of carefully chosen MOS sensors and the electronics is developed on a 16-bit embedded processor platform with low form factor and perfectly suited for continuous operations under field and factory environments. Once HEN is trained with sufficient number of known tea samples by an expert (human tea taster), the device becomes ready for assessing quality of given tea samples.

Determination of right fermentation end-point is very important to retain the quality of tea leaves during tea production. HEN can objectively do this by figuring out the occurrence of the second aroma peak (second nose).

Fig 14. Hand held E nose

Features

- A small, handheld, modular, aesthetic and easy-to-use instrument that mimics human perception of olfaction.
- Built on state-of-the-art low power embedded technology in conjunction with an array of MOS sensors.
- Two tea-aroma specific MOS sensors, developed by C-DAC's project partner, CGCRI are part of the sensor array.
- Low footprint optimized firmware to reduce memory and power requirements.
- The integrated odour handling unit is carefully designed and made simple.
- User interface through a 4.3-inch touch-screen display.
- Storage of training and test data in SD Memory card.
- Integrated rechargeable battery pack capable of 10 hours continuous operation.
- Rugged design to withstand harsh field and factory environments.
- Modular design for easy maintenance.
- Supervised training mechanism specific to user's choice.
- Can be trained with multiple training dataset corresponding to customer or region-specific choices in order to operate on any of them.

Extension Management Practices

i. **Initiatives:** The initiatives of Birsa Agricultural University are based on ICT including application of artificial intelligence (AI) and sensing technologies.

ii. **Reach:** The initiative being ICT based, the reach of agri-informatics services like technology portal, social media and Mobile App is worldwide. However, the contents in local languages are relevant only to the farmers and extension functionaries of Jharkhand state. The reach of community radio station is in the periphery of 10 km.As far as services of agri-electronics are concerned, the stakeholders are visiting the university.

iii. **Output:** The information available on technology portals, YouTube and Mobile App are accessed by the farmers and extension workers in pull mode based on their needs and interest (Table 2). Over 45,000 visitors have accessed information from technology portals comprising agriculture (33952), livestock (8344) and forestry (3772).

Table 3. Number of visitors of multilingual technology portals

S. No.	Particular	Number of visitors
1.	Multilingual Agriculture portal	33952
2.	Multilingual Livestock portal	8344
3.	Multilingual Forestry portal	3772

The analysis of viewership of "BAU EXTENSION" YouTube Channel reflects that farmers are preferring contents in local languages. The viewership of technology contents in Nagpuri language (Table 3) elucidates that farmers are preferring information on enterprise which have commercial value. The information on piggery has attracted viewership of about 16,000 followed by Ramdana (9752), Fennel, Ajoin, coriander and fenugreek (7025) and medicinal and aromatic plants (5788).

As far as agri-electronics equipments are concerned, farmers, officials and processors are visiting laboratory. About 500 interested clients from Jharkhand, Bihar and Bengal have visited the laboratory and have expressed their interests.

iv. **Outcome/Impact:** The farmers are slowly becoming IT friendly. The information accessed in pull mode from portal, YouTube and Mobile App shows that the initiatives have resulted in knowledge empowerment of the farmers. Farmers of the nearby areas, Government officials and representatives of farmer producer organizations (FPOs) and non-government organizations (NGOs) are regularly participating in radio program. As far as agri-electronics is concerned, a few rice millers have come forward to install E nose and E vision equipments.

Table 4. Viewership of top 10 Nagpuri technology contents on BAU EXTENSION

S. No.	Topic Name	URL	Viewership
1.	पशुधन उत्पादन प्रौद्योगिकी-पशुधन पालन -संकर सुवइर पालन	https://youtu.be/YEQ-z3PEp_0?si=U_kJBA5hMHX9yr7R	15799
2.	धान्य -रबी -रामदाना	https://youtu.be/61I-wzSEoo0?si=KC5BnyWrXtFOwBzx	9752
3.	सौंफ,अजवाइन,धनिया,मेथी	https://youtu.be/hsSS1lq1eW0?si=DZNUZ7E2DqFxFY6V	7025
4.	दवइवाला आउर सुगंधितिपउधामानक खेती-पामारोजा	https://youtu.be/553bCIVXJrs?si=tf4U-FiHMS-pLdsy	5788
5.	प्रसार शक्षिा	https://youtu.be/xIcUJ8iGShU?si=ucc7BqD7f_ZtWwWf	3876
6.	बीज उपचार	https://youtu.be/D6HvEq_NPd0?si=aTiMuhi8vm8_0M_q	3401
7.	दवइवाला आउर सुगंधितिपउधामानक खेती-सतावर	https://youtu.be/iUbUMp8J12I?si=gSJVUjDWpzyytkdR	3150
8.	अश्वगंधा,सर्पगंधा	https://youtu.be/ KDFKbRHQRkU?si=H120Q0IspWOm9I88	3036
9.	कालमेघ	https://youtu.be/_-gLxbMENnY?si=cWLsCXvkZ-KrzgmN	1753
10.	जैव प्रौद्योगिकी-उत्तक सवंद्धति (टशिू कल्चर्ड) पउधामानक खेती	https://youtu.be/FVjiJ947Nf8?si=fmdzp4XcSzjfOz9X	1520

Conclusion

The ever-evolving hardware and software technologies in Information and Communication Technologies (ICTs) are offering new opportunities and posing new challenges. ICT initiatives have been taken by numerous Government and Non-Government Organizations) including ICAR institutes and Agricultural Universities (AUs). Birsa Agricultural University, Ranchi has been proactive in the integration of ICTs in teaching, research and extension education. The initiatives of the university started with online portal, guided SMS, Interactive Voice Response System (IVRS) and offline system like video CDs and Learning Content Management System (LCMS). Considering the power of social media and community radio, the university launched YouTube channel (BAU EXTENSION) and established Community Radio Station (CRS). In next phase of its initiative, the university established Technology Resource Centre (TRC) on Agri-informatics and Agri-electronics consolidating earlier achievements and establishing agri-electronics equipments for the assessment of quality of agri-produce. Overwhelming response from the stakeholders has motivated us to go further in the integration of frontier technologies like Internet Of Things (IOTs), block chain and big data analytics.

References

1. Belattar, S., Abdoun, O., & Haimoudi, E. K. 2023. *Overview of Artificial Intelligence in Agriculture* (pp. 447–461). Springer International Publishing. https://doi.org/10.1007/978-3-031-43520-1_38
2. Bhat, P. P., R, R. P., K, A., Jadhav, A., N, M. K., M, R. C., & Reddy, S. L. 2024. The Role of Information and Communication Technology in Enhancing the Effectiveness of Agricultural extension Programs Worldwide: A review. *Journal of Scientific Research and Reports*, *30*(7), 963–976. https://doi.org/10.9734/jsrr/2024/v30i72206
3. Eli-Chukwu, N. 2019. Applications of Artificial Intelligence in Agriculture: A Review. *Engineering, Technology & Applied Science Research*. https://doi.org/10.48084/ETASR.2756
4. Fosu, A., & Giba-Fosu, N. 2024. The Concept of ICT on Agricultural Input Information: a Framework Design. *African Journal of Food, Agriculture, Nutrition and Development*, *24*(11), 24931–24946. https://doi.org/10.18697/ajfand.136.24135
5. Gupta, N., Gupta, P., Nadeem, D., A, A., & Elahi, A. 2023. Artificial Intelligence in Agriculture. *SSRN Electronic Journal*. https://doi.org/10.2139/ssrn.4345592
6. Jena, S. K. 2024. Digital Agriculture using advanced ICT and Agricultural Information Systems- The general and financial aspects in Indian Context. *Eonomic Affairs*, *69*(2). https://doi.org/10.46852/0424-2513.3.2024.37
7. Jha, K., Doshi, A., Patel, P., & Shah, M. 2019. A comprehensive review on automation in agriculture using artificial intelligence. *Artificial Intelligence in Agriculture*. https://doi.org/10.1016/J.AIIA.2019.05.004
8. Patel, K., & Patil, M. 2022. Artificial Intelligence in Agriculture. *International Journal for Research in Applied Science and Engineering Technology*. https://doi.org/10.22214/ijraset.2022.40308

9. Sahoo, P. K., & Sharma, D. 2023. Economic impact of artificial intelligence in the field of agriculture. *International Journal of Horticulture and Food Science*, *5*(1), 29–34. https://doi.org/10.33545/26631067.2023.v5.i1a.152
10. Shaikh, F., Memon, M., Mahoto, N., Zeadally, S., & Nebhen, J. 2022. Artificial Intelligence Best Practices in Smart Agriculture. *IEEE Micro*, 42, 17-24. https://doi.org/10.1109/MM.2021.3121279
11. Smith, M. 2020. Getting value from artificial intelligence in agriculture. *Animal Production Science*, 60, 46-54. https://doi.org/10.1071/AN18522
12 Svetskiy, A. V. 2022. Application of Artificial intelligence in Agriculture. *Sel'skoe Chozjajstvo*, *3*, 1–12. https://doi.org/10.7256/2453-8809.2022.3.39469
13. Yoe, H. 2024. *ICT-based Digital Agriculture*. 1. https://doi.org/10.1109/sera61261.2024.10685615.

5

Best Out of Waste: A Case Study on Banana Pseudostem

V.R. Naik[1], C.S. Desai[2] and Riddhi Trivedi[3]

[1]Research Scientist & Head, [2]Associate Research Scientist and [3]Senior Research Fellow, Soil and Water Management Research Unit, Navsari Agricultural University, Navsari

Email: vc@nau.in

Abstract

Secondary agriculture is a key to maximize that profit and major share can be gained from the agro waste utilization to create wealth. Banana (Musa sp.) is the second most important fruit crop in India next to mango. Its year round availability, affordability, varietal range, taste, nutritive and medicinal value makes it the favorite fruit among all classes of people. It has also good export potential. Banana is grown in about 120 countries. Total annual world production is estimated at 86 million tonnes of fruits. Apart from fruit yield, huge amount of waste biomass in the form of pseudostem, leaves, suckers etc., is generated. Disposal of such a huge biomass in unscientific way creates environmental problems. Using pseudostem, products like fibre, fabrics, paper, organic liquid nutrient, candy, vermi-compost, etc., have been developed. These products are capable of not only generating additional income for the farmers but can also provide alternative / supplemental eco-friendly natural raw material for some of the industries. NAU has been developed technologies for using sap in a profitable way. Sap can be used directly or through organic enrichment as liquid nutrient spray. Sap as liquid nutrient through drip was tested in most of the crops and results revealed that, about 10 to 15 per cent increase in yield of different crops was recorded. Being convinced of the beneficial effect of sap its enrichment process using only organic ingredients and incubating under anaerobic conditions was developed and standardized. In view of its novelty, the patent was obtained not only in India but also at international level. About 44 MoU is done with the private parties for commercial production and marketing of Organic Liquid nutrient and Central core candy. This Novel organic liquid nutrient is now available throughout India with different brand names by private companies.

Keywords: *Banana Pseudostem, Nutrient, Novel, Organic*

Introduction

Banana is a common man's fruit grown in about seven lakh ha area in India. Apart from fruit, banana crop generate huge quantity of waste biomass in the form of pseudostem, leaves, suckers, etc. This biomass is absolute waste in most of the states of India as of now. Further farmers are spending about Rs. 15000 to 20000/ha for disposing off this waste from their fields. The present mode of disposing off pseudostem waste in nallas, borrow pits, road side, burning *etc.,* which are causing environmental problems. In order to utilize this waste particularly pseudostem for developing different products, a project *titled;* "A Value Chain on Utilization of Banana Pseudostem for Fibre and Other Value Added Products" was sanctioned under NAIP in consortium mode by ICAR, New Delhi during 2008.

The project entitled, "Establishment of Secondary Agriculture unit for skill development in students and farmers" was sanctioned by PIU, under the aegis of NAHEP-CAAST sub project in 2018. Under this project "Value addition and waste management of horticultural produces" was examined in special reference to banana pseudostem as supplemental raw material for NOVEL pesticide and fungicide.

In view of this, Soil and Water Management Research Unit, N.A.U., Navsari have worked in consortium comprises of Central Institute for Research on Cotton Technology (CIRCOT), ICAR, Mumbai, Manmade Textile Research Association (MANTRA), Surat and J. K. Mills Ltd., Songadh as the partners. The basic aim of this project was to develop different value added products using presently waste banana pseudostem as raw material.

As envisaged in project proposal, the value added products *viz;* Fibre and non-woven fabrics, handmade papers and boards, vermi-compost, enriched sap (NAUROJI Novel organic liquid nutrient) and candy have been developed which are techno economically viable. The viability is substantiated by the fact that more than ten private parties have already signed MoU with NAU, Navsari for commercial production and marketing of organic liquid nutrient prepared using banana pseudostem sap and central core candy. Simultaneously, processes for the products like microcrystalline cellulose from Fibre, sap as a mordant in textile dying, pseudostem scutcher based vermicompost as fish feed, etc. have also been standardized during the current project period. This has become possible due to highly dedicated efforts put in by the team of scientists from all the four partners and full moral boosting supports by their authorities in implementing the project smoothly.

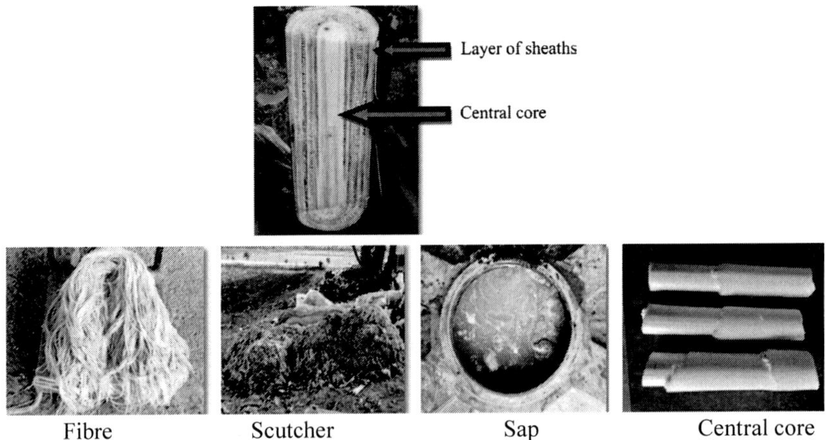

Fig 1. Value chain on utilization of Banana Pseudostem for Fibre and other value added products

Fibre-based Products

Banana Fibre is eco friendly and chemical free. It is grease proof, water and fire resistant and totally bio-degradable. It is the fibre extracted from the trunk of Banana tree which is considered as a waste. Banana Fibre is used in manufacturing industries of fabric woven as well as non-woven, Microcrystalline Cellulose (MCC), handicrafts, home decorative, door mats, table mats, pooja and meditation mats. Banana Fibre has got very wide usage in the paper units *like*, 100 per cent chemical free tissue paper, filter paper, paper bags, craft papers, carry bags, decorative papers, bond papers, products like pen stands, table decorative, land shades *etc.*, Products that are made out of banana Fibre has very good market.

Scutcher-based Products

1. Vermicompost

Huge quantity of scutcher (about 30 to 35 t/ha) is generated during fibre extraction. The process has been standardized for vermicompost preparation using pseudostemscutcher and dungs. The vermicompost prepared had been tested for its quality and is being marketed in the NAU trade name NAUROJI. Application of vermicompost @ 3 kg /plant in banana and 5 t/ha in sugarcane in addition to recommended dose of fertilizer recorded comparable yields of both the crops with FYM and biocompost. Use of vermicompostalso sustains soil health.

2. Fish Feed

An innovative experiment was conducted to explore the feasibility of blending vermicompost with fish feed. The results of two-year study revealed that the routine fish feed (cattle feed) can be substituted by vermicompost up to 30 per cent without any reduction in body weight of fish.

3. Sap

Organic Liquid Nutrient: About 12,000 to 15,000 litres of sap can be obtained from the pseudostem obtained from one hectare. In addition to the direct use of sap as liquid fertilizer, its enrichment process has been standardized. That product launched by NAU in known as "NAUROJI Novel Organic Liquid Fertilizer". Apart from essential plant nutrient, it also contains growth promoting substances *viz.*, GA and cytokinin. Enriched sap of about 2500 litres have been prepared and distributed among the farmers for demonstration. The enriched sap has been tested through foliar application in different crops. Foliar application of enriched sap could improve the vigour of brinjal and chilli nursery. The seedlings become ready for transplanting 6 to 8 days earlier than traditionally grown. Similarly, fruit retention in mango was increased by 50 per cent due to 4 sprays of enriched sap.

4. Central Core

Central core is inner most tender portion of the pseudostem which is edible. About 10 to 12 t/ha central core can be obtained. The work has been taken up for standardizing processes for developing various edible products from it. Process for preparing candy, RTS and pickles has been standardized. Left-out syrup after preparing candy can be used for preparing RTS. This RTS can be flavoured with any natural or synthetic flavours.

Various By-products Prepared from Banana Pseudostem

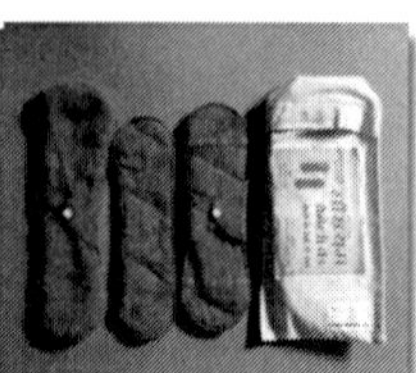

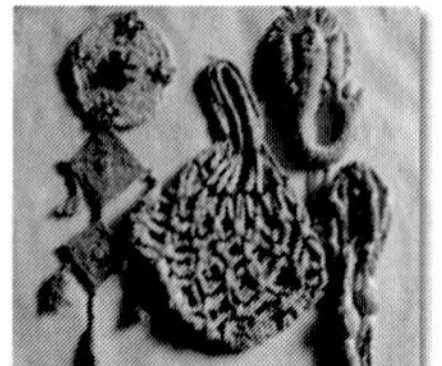

Fig 2. Fabrics, Sanitary napkins, Handicrafts and non-woven sheets made from Banana Fibre and Yarn

Fig 3. Handmade papers, files, folders, currency grade papers made from banana Fibre

Fig 4. Vermicompost from Scuture

Fig 5. Fish Feed from Scuture

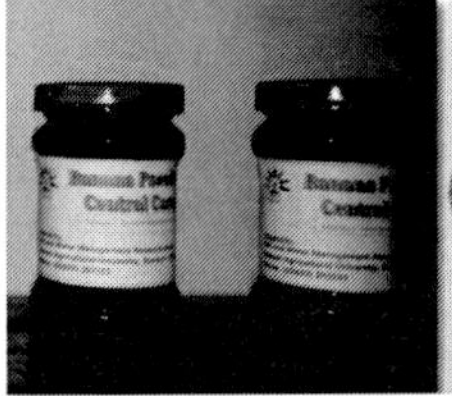

Fig 6. Candy, chutney, jam, jelly, cookies (edible products) prepared from Central Core

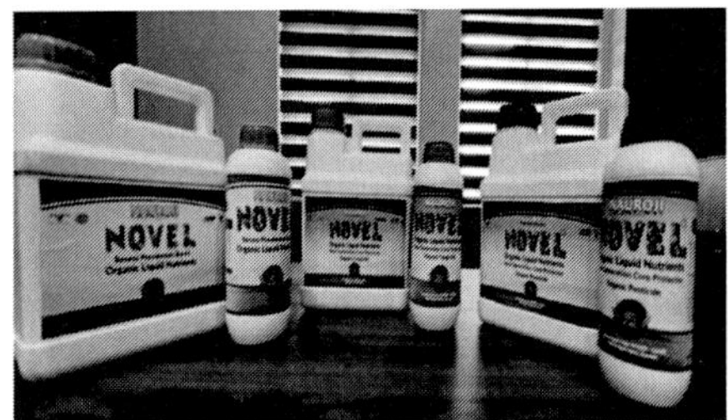

Fig 7. Organic Liquid Nutrient, pesticide (plus) and fungicide (prime) prepared from Banana

About the Technology Addressed and Extension Management Practices

Initiative: Preparation and evaluation of enriched sap (Organic Liquid Nutrient) and scutching waste based vermicompost. Banana Pseudostem Sap (BPS) has been extracted from the outer sheath of pseudostem of banana tree. It looks like colourless clean water immediately after extraction. However, with the passage of time, it slowly turns into a light khaki colour due to the oxidation of phenolic rings present in it. Banana pseudostem sap is obtained as a by-product during extraction of fibre. It is a rich source of plant nutrients like K, Fe and plant growth regulators. Hence, experiments were conducted to use the sap directly as a liquid nutrient initially in different crops through soil and foliar application. In response to its positive effect in enhancing the yields of the crops, enrichment process was standardized and tested in different crops at NAU, Navsari.

Use of Sap as a Liquid Nutrient (without enrichment)

Separation of Sap from Banana Pseudostem Scutcher: From the scutcher + sap collected in tray placed bellow raspador machine during fibre extraction, sap can be separated either by squeezing the scutcher manually or by using screw press developed by NAU, Navsari. Use of screw press is more efficient in sap separation from scutcher than manual squeezing. On an average, from one-hectare banana plantation around 12,000 to 15,000 liters sap is obtained. In other words, one unit of five raspador machines can generate about 4000 to 5000 liters sap per day.

Fig 8. Hydraulic press m/c for extracting sap from scutcher

Table 1. Nutritional and biochemical composition of banana pseudostem sap (on fresh weight basis)

Parameter	Unit	Content (range)	Mean
N	%	0.005 – 0.032	0.015
P		0.002 – 0.007	0.0028
K		0.154 – 0.234	0.208
S		Trace – 0.004	0.002
Ca		0.004 – 0.020	0.014
Mg		0.014 – 0.101	0.048
Fe	ppm	2.57 – 38.17	10.72
Mn		2.66 – 19.34	9.78
Zn		Trace – 1.79	1.07
Cu		Trace – 4.30	0.98
Total soluble sugars	(mg/ml)	0.356 - 4.881	1.877
Total phenols	(µg/ml)	2.750 - 25.19	13.803
Total amino acids	(mg/ml)	0.022 - 0.232	0.129
Urease activity	(U/ml/min)	1.450 - 10.14	4.676
Cytokinin (mg/l)	(mg/l)	44.5	
Gibberellic acid	(mg/l)	13.7	

The range and mean values reported in table revealed that variation in the composition of sap vary with the element which is higher in N, Cu and biochemical parameters. Based on the mean value K, Fe, Zn and Mn are higher as compared to rest of the elements. Among the biochemical parameters, phenol content recorded higher value in comparison to rest of the parameters. The sap also contains plant growth regulators in appreciable quantity.

A. Development of Process for Preparing Enriched Sap

(NAUROJI Novel Organic Liquid Nutrients, Novel Plus & Novel Prime)

Process: The banana pseudostem sap is collected by squeezing scutcher waste either manually or by press obtained during the process of fibre extraction. The sap obtained is to be filtered using muslin cloth for removing the suspended material. Mixing of different organic inputs and sap has to be done in sequential manner (Patent, PCTIB2012053268, 1609/ MUM/2011).

Fig 9. Novel Organic Nutrient Liquid

Fig 10. Novel Plus - Organic Liquid Nutrient with pesticide

Fig 11. Novel Organic Liquid Nutrient with fungicide

The whole mixture is then filled in bio-digester and incubated under anaerobic condition. The mixture is to be stirred periodically. After specified period the supernatant is to be collected, filtered and stored in air tight container.

B. Composition of Enriched Sap

In order to know the variation in composition of sap during anaerobic decomposition, periodical samples were collected and analyzed for major and micronutrients content by adopting standard analytical procedures. The results indicated no variation in content except N, P, Fe and Mn which tended to increase with the time of incubation. Similarly, after completion of incubation, ready for use enriched sap (OLN: organic liquid nutrient) was analyzed for nutrient, biochemical and microbial parameters. Based on the composition of enriched sap, it can be used as spray solution in different crops. It contains not only essential plant nutrient but also plant growth regulators *viz.* cytokinin and GA_3 as well as some beneficial organisms.

Table 2. Composition of enriched sap ready to use

Chemical			Biochemical		
Parameters	**Unit**	**Mean**	**Parameters**	**Unit**	**Content**
N	%	0.062	Total phenol	mg/100 ml	48.0 – 49.1
P		0.018	Urease activity	U/ml/min	63 – 81
K		0.180	Gibberellic Acid	mg/l	110.2 – 205.0
Ca		0.031	Cytokinin	mg/l	137.8 – 244.3
Mg		0.092	**Microbes**	**Unit**	**Population**
S		0.010	Total viable count	(CFU/ml)	1065 x 10^3
Mn	ppm	5.73	PSB		1025 x 10^2
Cu		0.40	Rhizobium		285 x 10^2
Zn		2.92	Azotobacter		460 x 10^2
Fe		109.3	Fungal count		1200

Output

Nauroji Novel Organic Liquid Nutrients: Which is aimed at cultivating the land and raising crops in such a way, as to keep the soil alive and in good health by using banana pseudostem based wastes (sap) and other biological materials along with beneficial microbes (biofertilizers) to release nutrients to crops for increased production in an eco-friendly pollution free environment. With the addition of botanical substitutes, farmers have been able to produce bigger crops on less land, increasing crop productivity by 20 to 50 percent.

The products contains not only essential plant nutrient but also plant growth regulators *viz.* NAA, cytokinin and GA3 as well as some beneficial soil conditioning as well as waste decomposing organisms. It also contains bacteria which can improve soil health and can be useful in different stages of plant growth *e.g.* vegetative development, flowering, fruit setting, fruit development *etc.* It can be used in different crops in different stages by various methods like fertigation, drenching, foliar spray, injection, cone feeding *etc.*

NOVEL: Organic Liquid Nutrients

- Pure organic product
- Supply plant nutrients (N,P,K) including micronutrients
- Contains naturally occurring growth promototers
- Improves soil physical properties like structure, water holding capacity etc.
- Contains botanical pesticides
- Increases the availability of nutrients
- Used in agriculture to improve soil fertility and enhance crop productivity.

- Enhance vigorous root development and growth
- Enhances total plant growth, number of flowers, maximum conversion of flowers into fruits and pods.
- Growth promoter; it is very helpful in nursery plants.
- Reduces fruit drop and increase fruit size and setting.

NOVEL PRIME (Fungicide)

- NOVEL Prime (fungicides) refer to products that derived from natural plant sources with no chemical alteration.
- Organic supplements with botanical fungicides.
- Contains major/micro nutrients.
- Contains naturally occurring PGRs
- Fortified botanical fungicides can control most of the anthracnose, mildews, leaf spot, rust, blasts, wilting and other harmful fungi/virus/bacteria.
- Can give plant a complete food
- Improves plant health and its vitality because of availability of plant nutrients and hormones, which ultimately resulting in increasing crop yield and reducing cost of production.

NOVEL PLUS (Pesticide)

- It refers to products that derived from natural plant sources with no chemical alteration
- It is organic supplements with botanical pesticides.
- Can give plant a complete food
- Contains major/micro nutrients.
- Contains naturally occurring PGRs
- Fortified botanical pesticide can control most of the insect-pest like aphids, jassids, white fly, caterpillars, borers and other harmful insect/pests.
- It can improve plant health and its vitality because of availability of plant nutrients and hormones, which ultimately resulting in increasing crop yield and reducing cost of production.

Outcome and Impact

The liquid portion sap obtained along with scutcher during fibre extraction from pseudostem by raspador machine is good source of plant nutrient along with growth promoting substances like cytokine, GA_3, etc. If used as liquid nutrient either through drip system or drenching in crops like banana, sugarcane, papaya, onion, leafy vegetable, it can save 20 to 40 per cent dose of nutrient with yield advantage of 10-15 per cent. This can be directly adopted by the farmers themselves.

Apart from direct use of sap as liquid nutrient, an enrichment process was developed (patented) for preparing Novel Organic Liquid Nutrient (OLN) suitable for foliar and soil application. It was tested in mango, banana, wheat and paddy crops. Application of 3 – 4 sprays of OLN @ 1 to 2 % (v/v), could increase the yield by 12 to 15 per cent across the crops. The OLN has been prepared using only organic inputs and hence suitable for use in organic farming system as liquid formulation.

Fig 12. Photographs depicting impacts of product

- These are some photographs of various crops shared by farmers with us, after using NOVEL organic liquid nutrients. They have good experience with these NOVEL products and also increased their yield and quality of produce.
- For the benefit of the farmers of globe and to make it available in every corner of earth, Navsari Agricultural University has done agreements through technology transfer for commercial scale production and Marketing of it with different private companies.
- Currently, it is available in the commercial market with different brand names and all brand packaging contain NAU symbol on it as a symbol of trust and quality.
- MoU signed with forty-two (42) National and two (02) International firms for commercial production and marketing of organic liquid nutrient from banana pseudostem sap and candy from central core.
- NAU has commercially sold out more than 2 lakh litres organic liquid nutrient; 17,478 litres novel plus and 3123 litres novel prime since year 2019 to 2024 to the farmers.

- Standard recommendation of organic liquid nutrient by NAU for individual crops is 1-2%. Farmers are advised to use it as per the suggestion of NAU.

Table 3. Selling data of OLN from last five years

Year	Novel Organic Liquid Nutrient	
	Total sold out (liter)	**Total Revenue (INR)**
2019-20	19,285	23,74,590/-
2020-21	29,629	36,24,650/-
2021-22	19,090	25,29,770/-
2022-23	18,072	24,02,260/-
2023-24	16,289	28,63,400/-
	102,365	

Conclusion

From environmental and social safeguard point of view, all the activities of this project have scored positive points. Not only this, but this project has capacity to generate rural employment (183 man-days/ha annually), provide alternative/supplemental eco-friendly raw material to textile, paper, pharmaceutical, nutrient and confectionary industries. A farmers growing banana can realize an additional income of Rs. 60,000 to 70,000 per ha from fibre, sap and vermicompost preparation on annual basis. In a nut shell, this project has shown a new path for generating wealth from waste in a most eco-friendly way and the income sharing among the farmers, entrepreneurs, industries and end users as well.

The scaling up activity needs to be taken up in comprehensive way on cluster basis. Further, government/entrepreneurs/ farmers' cooperative should act as a facilitator between banana growers and industry for confidence building between them. In this direction, National Horticulture Mission, MoA, GoI, New Delhi has already uploaded the information of this project on its website. In order to give insight in the monetary benefits in pseudostem based products, business models for fibre extraction and preparing non-woven, preparing organic liquid nutrients from pseudostem sap and preparing candy from central core of banana have also been included in the report. Apart from this, the status of banana pseudostem processing before and after NAIP interventions is also reported diagrammatically. Properly dealing with discarded products can reduce the potential for environmental pollution while also protecting the individual who is responsible for the discarded materials.

References*

1. Desai C. S. Desai; Patel J. M. and Pawar S. L. 2015. Training manual for processing of banana pseudostem to value added products. NAU, Navsari.
2. Desai C. S.; Desai S. K.; Desai. C, D.; Mistry P. S. and Vaidya H. B. 2015. Effect of pre harvest production technology of banana on qualitative evaluation of various processed products. Trends in bioscience. 8(22), 6165-6177.
3. Desai C. S.; Desai S. K.; Desai C. D. and vaidya H. B. 2014. Kel ek kalpataru. SWMRU, NAU, Navsari
4. Desai G. B.; Anand V; Sonvene S. S. and Patel J. M. (2010-11) Effect of enriched banana pseudostem sap at pre flowering stage on production and quality of banana *var.* G. Naine. AGRESCO report NAU, Navsari
5. Desai. C. S.; Patel J. M.; Pawar S. L.; Usadadia V. P.; Naik V. R. and Savani, N. G. 2016. Book Value added product from banana pseudostem.
6. Kolambe, B. N.; Desai. S.; Patel K. K.; Patel P. S.; Desai S. K. Pawar, S. L.; Patel J. M. and Patil R. G. 2013. Value added products from banana fruits and pseudostem. National Seminar on tropical and sub-tropical fruits. 9-11.
7. Pandit P. S.; Shukla S. P.; Desai C. S and Vaidya H. B. (2014) Preparation of Ready to serve beverage from banana pseudostem sap. AGRESCO report NAU, Navsari.

* The content of this chapter was prepared in consultation with above references.

6

Biofertilizers: An Initiative to Eco-Friendly Supplement the Plant Nutrients and Protect Environment

Lalit Mahatma[1*], M.I. Patel[2], K.B. Rakholiya[3], T.R. Ahlawat[4] and Z.P. Patel[5]

[1]Professor of Plant Pathology and Associate Director of Research, [2]Assistant Professor, Department of Plant Pathology, [3]Professor & Head, Department of Plant Pathology, [4]Director of Research & Dean PG Studies, [5]Vice Chancellor, Navsari Agricultural University, Navsari

Email: mahatmalalit@yahoo.co.in

Abstract

Biofertilizers produced by Navsari Agricultural University, Navsari was rapidly accepted, adopted and became popular among the farming community of South Gujarat because of its high quality and extensive extension services. These biofertilizers are being produced and supplied in the brand name NAUROJI. Different biofertilizers are Nitrogen fixating (Azotobacter chroococcum, Rhizobium spp., Gluconacetobacter diazotrophicus. Azospirillum brasilense, Azotobacter chroococcum) Phosphate Solubilizing Bacteria (Bacillus spp. and Pseudomonas spp.) and Potash Mobilizing Bacteria (Frateuria aurantia). Nitrogen fixing bacteria fix atmospheric nitrogen and give it to plants. Whereas other bacteria produce organic acid which help in enhancing the availability of the essential nutrition by converting unavailable form of nutrients into available form. They have potential to save 50 per cent chemical fertilizers including nitrogen, phosphorous and potash. In last fifteen years the University has produced and supplied more than 10 lakh liters of biofertilizers and PGPR to the farmers at very nominal price. It was estimated that these biofertilizers covered 66,600 ha land area and saved use of 295.7 tonnes urea. According to the criteria of IPCC, by saving 295.7 tonnes of urea, 59.14 tonnes CO_2 emission has been reduced. Similarly, use of the phosphate solubilizing bacteria and potash mobilizing bacteria also have saved considerable CO_2 emission and helped in protection of environment.

Keywords: *Azospirillum brasilense, Azotobacter chroococcum, Azotobacter chroococcum, Bacillus spp., Biofertilizers, CO_2 emission, Frateuria aurantia, Gluconacetobacter diazotrophicus. IPCC, Navsari, Pseudomonas spp., Rhizobium spp.*

Introduction

In 1804, Nicolas Theodore de Saussure demonstrated that plants don't uptake carbon from the humus, rather it comes from the carbon dioxide in the atmosphere. He further stated that soil is source of plant nutrients. Subsequently in 1840, Carl Sprengel proposed law of minimum which mean the growth is regulated by scarcest (minimum available) resource instead of total resources. During the same time, Justus von Liebig observed that the chemical elements nitrogen (N), phosphorus (P), and potassium (K) are essential for the plant growth. He strongly popularized the theory of mineral nutrients and law of minimum by developing the barrel with multiple holes at different height and filling water in it. It has opened the door of fertilizers industries; therefore, Justus von Liebig has been described as the "father of the fertilizer industry".

Consequently, many fertilizer companies started producing various chemical fertilizers all over the world. This had increased the productivity of the various crops. In India the first fertilizer factory which produced Single Super Phosphate (SSP) was started in Ranipet, Tamil Nadu in 1906 (Chaudhary, 2023). Before, green revolution we have established fertilizers industries of Ammonium Sulphate (1933), Ammonium Sulphate Nitrate (1959), Urea (1959), Ammonium Chloride (1959), Ammonium Phosphate (1960), Calcium Ammonium Nitrate (1961), Nitro phosphate (1965) and Di Ammonium Phosphate (1967). The green revolution ushered in 1967-68 was due to introduction of input responsive high yielding dwarf variety of wheat and use of chemical (mineral) fertilizers.

These chemical fertilizers alone were responsible for increasing the yield of wheat by 233 per cent (Khunt et al. 2014&2015). If chemical fertilizers would not have been used for the crop production, we would not be in the position to feed the burgeoning human population. In the pursuit of increasing productivity, we injudiciously used chemical fertilizers and overlooked the deleterious effect of it on animal, human and environmental health. Production and use of chemical fertilizers, has increased the level of Green House Gases (GHGs) in the atmosphere. To reduce GHGs and their environmental impact, alternate source of plant nutrients *viz*., Farm Yard Manures (FYM), composts, organic waste-derived amendments, biofertilizers, etc. started gaining popularity. Organics are also considered unsafe as during its degradation it produces CH_4. Further FYM is limitedly available. Contrary, biofertilizers don't produce any greenhouse gases and can be prepared in the laboratory to fulfill any growing demand. Potential of microbes (Biofertilizers) in enhancing the availability of the essential nutrition by fixing the atmospheric nitrogen and converting unavailable form of nutrients into available form was known to us since long. However, glare of the chemical fertilizers, its business potential and easy availability for crop production has limited the efforts to popularize biofertilizers. Therefore, it is the right time to popularise the use of the biofertilizers technology and protect our environment without compromising the health of animals and humans. Realizing its significance, Ministry of Agriculture, Department of Agriculture and Cooperation, Government of India, New Delhi, vide their order of dated March 24, 2006 included biofertilizers and organic fertilizers under section 3 of the Essential Commodities Act, 1955 (10 of 1955), in Fertilizer (Control) Order, 1985. These rules were further amended in respect of applicability, specifications and testing protocols vide Gazette notification November 03, 2009. Subsequent notifications have further expanded the number of products for use in agriculture. This has opened the doors of biofertilizers industries which have high significance in the existing agricultural scenario.

About the Technology Addressed

Biofertilizers have potential to save 50 per cent chemical fertilizers including nitrogen, phosphorous and potash (Mahatma *et al.*, 2016a & b). Nitrogen is available abundantly (78 per cent) in the atmosphere as inert nitrogen (N_2) gas, whereas plant uptake it in the form of nitrate, ammonium ions, and available amino acids from organic sources (Zayed *et al.*, 2023). Phosphorous is the second-most important nutrient of plants, absorbed as $H_2PO_4^-$ in acidic pH or as HPO_4^{2-} in alkaline pH (Sabalpara and Mahatma 2019). After the application of phosphorous fertilizers in soil, chemical reactions convert available form of phosphorous in insoluble and unavailable forms by forming a strong covalent bond. In alkaline soils, calcium is the dominant cation that reacts with phosphate. A general sequence of reactions in alkaline soils is the formation of dibasic calcium phosphate dihydrate, octacalcium phosphate, and hydroxyapatite. The formation of each product results in a decrease in phosphate solubility and availability. When pH is acidic, plant-available phosphorus becomes increasingly tied up with aluminum as aluminum phosphates (Sabalpara and Mahatma 2019).

Potassium is the third most important nutrient for plants and the fourth most abundant nutrient, constituting about 2.5% of the lithosphere (Sabalpara and Mahatma 2019). Scenario of potash is entirely different and available in the soil in four different forms viz., water soluble, exchangeable, non-exchangeable and mineral form. These different forms are not homogeneously distributed, rather they are in dynamic equilibrium with each other. This potassium dynamic in soil is based on the magnitude of equilibrium among various forms of potassium and generally controlled by the physicochemical properties of soil. About 98% of total potash is in the form of primary (micas and feldspars) and secondary (illite group) clay minerals form which can't be utilized by the plants. Water soluble and exchangeable potash are only available to plants (Harpreet Kaur 2019). Deficiency of the one nutrient can only be compensated by the same nutrient. irrespective of availability other nutrients in any quantity. Theory of conservation of mass exists in the natural ecosystem (Mahatma *et al.*, 2023). None of the essential nutrient can be synthesized in the laboratory. In the process of fertilizer synthesis, we use or mine these unavailable from of nutrients from the natural sources and convert them in concentrated available from through the series of chemical reactions. To enhance the availability of these nutrients where it is required, the fertilizers are applied.

Different Microbes used as Biofertilizers

Microorganism fixes atmospheric nitrogen and provides it to the root zone in the form assimilated by the plants. They can also convert the unavailable form of the nutrients in the available form and can be considered as micro-industries for the fertilizer synthesis. Fertilizers synthesis by these microbes are eco-friendly and economic. Few microbes can fix atmospheric nitrogen either symbiotically (*Rhizobium* spp.), associative symbiotically (*Azospirillum brasilense* and *Gluconacetobacter diazotrophicus*) or living freely in the root zone (*Azotobacter chroococcum*). These can be good source of nitrogenous fertilizers (Sabalpara and Mahatma, 2016 and 2017). Members of the leguminous family evolved a specific mechanism to overcome the nitrogen deficiency by establishing symbiotic relationship with the nitrogen fixating bacteria and most studied symbiotic relationship (Mahatma *et al.,* 2019).

There are many microbes (*Bacillus* spp., *Burkholderia* spp. and *Pseudomonas* spp.) which release series of organic acids and break the covalent bond between the phosphate and cations. In the process the phosphorous is released in the form of $H_2PO_4^-$ or HPO_4^{2-} which is assimilated by plants. Similarly, there are certain microorganisms (*Acidothiobacillus ferrooxidans*, *Bacillus mucilaginosus*, *Frateuria aurantia*) that release various metabolites which enhance rate of weathering of the minerals and rapidly convert non-exchangeable form of potash in to water soluble and exchangeable form of potash. These microorganisms if used in an efficient manner can serve as an excellent source of nutrients for the plants. Apart from these, *Pseudomonas* is one of the widely recognized species for the production of lipase (Jaeger et al,

1994; Kojima *et al.*, 2003). The lipase produced by these species is of Group I type which is composed of approximately 285 amino acid and molecular weight 30 KD. *P. pseudoalcaligenes* is also able to use cyanide as a nitrogen source, and as a result, it is widely used for bioremediation (Anzai *et al.,* 2000, Topiwala and Mahatma, 2017). These biofertilizers can be applied in the soil directly or through drip irrigation also for seed treatment, root dip treatment, sett treatment, field application, spot application.

Extension Management Practices

Initiative

Navsari Agricultural University (NAU), Navsari initiated work on different aspects of biofertilizers since its inception in 1965. In 2006 an Experiential Learning Programme on Biofertilizers was sanctioned by ICAR, New Delhi and facilities were created. The project started functional from May 2008. Spectacular research findings and extension efforts were made in the biofertilizers in 2009 by the NAU, Navsari. Series of lab and field trials were performed and mass multiplication protocol of different biofertilizers standardized. License for the commercial production of the biofertilizers was also obtained as per the Fertilizer Control Order (FCO)-1985. The brand name of the products being prepared by NAU, Navsari is NAUROJI (Fig 1). The name has been coined to pay tribute to able son of Navsari Shree Dadabhai Naoroji who also known as the "Grand Old Man of India". The three letters of the brand name also represent the name of Navsari Agricultural University.

NAU, Navsari isolated potential microbes viz., *Azotobacter chroococcum, Rhizobium* spp. (*Rhizobium, Bradyrhizobium, Sinorhizobium, Mesorhizobium, Allorhizobium*, etc), *Gluconacetobacter diazotrophicus. Azospirillum brasilense, Frateuria aurantia, Bacillus* spp., and *Pseudomonas* spp., etc. Not only bacteria, a fungi *Penicillium expansum* NAUG-B1having phosphate solubilizing activities was also isolated and thoroughly characterized. HPLC analysis detected gluconic acid as major organic acid in the course of phosphate solubilizing (Panchal *et al.*, 2015). Different isolates were characterized and selected most efficient microbes for biofertilizers production. For the identification of most efficient strain, experiments were perfumed *in vitro*, in pots and finally in field on different crops (Fig 2).

In biofertilizers production and supply, NAU, Navsari has succeeded in winning the confidence of farmers by overcoming all the technical and administrative hurdles within a period of short span. Farmers can apply these biofertilizers in all the field, horticultural and plantation crops. Research work carried out at NAU, Navsari proved that biofertilizers have potential to reduce the use of chemical fertilizers to the tune of 50 per cent and increase the productivity upto 20 per cent. The microorganisms in biofertilizers restore the soil's natural nutrient cycle and build soil organic matter (Mahatma *et al.*, 2016a & b). Not only the major nutrients, but the minor nutrients are also made available to the plants, thereby, maintain overall nutrient balance in the crop. Application of biofertilizers and subsequent amendment of organic matter for the long duration help in establishment of these microorganisms in the soil ecosystem. These microorganisms may be endophytic or colonize near the rhozosphere and also act as PGPR to helps plant stand better in various biotic and abiotic stresses.

Reach

To reach to the farmer extensively in rapidly changing agricultural situation, Government of Gujarat (GoG) has introduced a novel approach, Krishi Mahotsava. Prima facie, the approach was to bridge the gap between the technology and the farmers. This approach expedites the process of technology diffusion in the different parts of the state which was aimed to move towards demand-led approach to extension services for the farmers. In the Krishi Mahotsava program, kits of innovative novel products were distributed to the farmers. Scientists also have paid attention in sensitizing the famers about these innovative technologies.

These biofertilizers developed by the NAU, Navsari were included in the kits which were used by farmers in their fields. The product was very useful and inquiry about the product increased. For further information, farmers contacted Department of Plant Pathology, where biofertilizers are being prepared. The department provided detailed information to the farmers about various biofertilizers and encouraged them to use in their fields. Through mouth to mouth extension, demonstration, training by the KVKs and departments, the demand of the product has increased significantly. Without much efforts of marketing the product is being obtained and applied by the farmers (Fig 3). Krishi Mahotsava aimed at preparing micro level plans covering each block and village, therefore, the technology reach every village. In a series of video conferences, the then Hon'ble Chief Minister of Gujarat Shree Narendra Modi had addressed farmer's meetings organized during Krishi Mahotsava. He had a special program on biofertilizers which further helped in disseminating the technology.

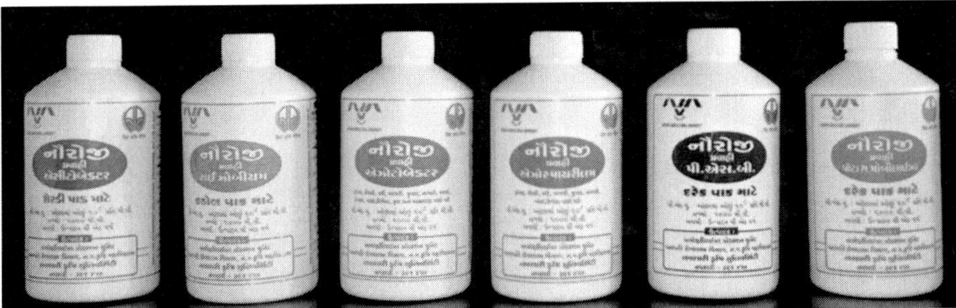

Fig 1. Packing of different products of biofertilizers being prepared at NAU, Navsari

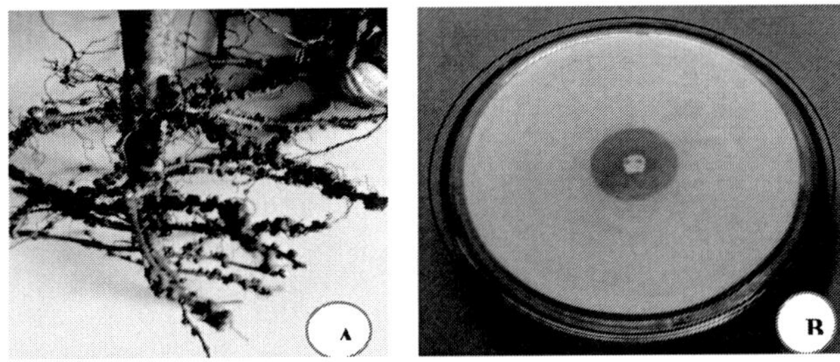

Fig 2. (A) Groundnut plant showing Rhizobium nodules (B) Quality of PSB as tested in the laboratory as per FCO-1985 norms

Fig 3. Field view of chilli crop with 50 per cent reduced chemical fertilizers and biofertilizers (Azotobacter, PSB & KMB)

Output

Due to quality product and massive extension activities, the demand of the biofertilizers has increased in the Gujarat. The University signed a MoU with the Shree Mahuva Pradesh Sahakari Khand Udyog Mandli Ltd., Bamania, Ta. Mahuva, Dist. Surat in 2012 for the preparation of biofertilizers and distribution to their farmer members. Apart from these, the Department of Plant Pathology, N.M. College of Agriculture, NAU, Navsari is preparing various biofertilizers which is being used by the farmers. In the last fifteen years the department has produced and supplied more than 10 lakh liters of biofertilizers and PGPR to the farmers at very nominal price.

Outcome/ Impact

Impact of the technology is multifold. Biofertilizers are economic and eco-friendly way of providing plant nutrients. It has reduced the dependency of chemical which are harmful to the animal, human and environmental health. Total 10 lakh liters of biofertilizers were used by the farmers in the South Gujarat. Among this, the nitrogenous biofertilizers was one third (approx 3.33 lakh liters) of the total biofertilizers. The doses of biofertilizers in the different crops varied from 2.5 liters to 5.0 liters per hectare (Mahatma *et al.*, 2016a &b). This 3.33 lakh liters biofertilizers if applied uniformly at 5.0 liters per hectare, would have covered 66,600 ha land area. Generally, paddy and sugarcane are grown in the South Gujarat which required 100 and 250 kg nitrogen, respectively (Kumari *et al.*, 2025; Mahatma *et al.*, 2016a &b).

Even if minimum 100 kg nitrogen is applied to one-hectare land than to supply this minimum recommended doses, farmer need to apply 222 kg urea. Accordingly, to cover 66,600 ha area, farmers need to apply 14,78.5 tonnes urea. Biofertilizers have potential to reduce the use of 50 per cent chemical fertilizers and increase 20 per cent yield. Even if 20 per cent chemical fertilizers is saved by the biofertilizers, then by applying 3.33 lakh liters of nitrogen fixing biofertilizers, farmers have saved 295.7 tonnes urea. The IPCC proposed that the CO_2 emission factor from urea is 0.2 kg C per kg urea (Kim *et al.*, 2017). Accordingly, by saving 295.7 tonnes of urea, 59.14 tonnes CO2 emission has been reduced by the use of biofertilizers. Similarly, use of the phosphate solubilizing bacteria and potash mobilizing bacteria also have saved considerable CO_2 emission and helped in protection of environment.

Conclusion

Chemical fertilizers alone were responsible for increasing the yield of wheat by 233 per cent during the green revolution. In the pursuit of increasing productivity, injudicious use of chemical fertilizers resulted in the deleterious effect on animal, human and environmental health. Production and use of chemical fertilizers, has increased the level of Green House Gases (GHGs) in the atmosphere. To reduce GHGs and their environmental impact, alternate source of plant nutrients microbes (biofertilizers) are the best alternate. Biofertilizers help in enhancing the availability of the essential nutrition by fixing the atmospheric nitrogen and converting unavailable form of nutrients into available form was known to us since long. They have potential to save 50 per cent chemical fertilizers including nitrogen, phosphorous and potash.

One can't synthesize any of the nutrient, however, in the process of fertilizer synthesis one use or mine these unavailable from of nutrients from the natural sources and convert them in concentrated available from through the series of chemical reactions.

Navsari Agricultural University (NAU), Navsari initiated work on different biofertilizers viz., Nitrogen fixating (*Azotobacter chroococcum, Rhizobium* spp., *Gluconacetobacter diazotrophicus. Azospirillum brasilense, Azotobacter chroococcum*) Phosphate solubilizing (*Bacillus* spp. and *Pseudomonas* spp.) and Potash mobilizing (*Frateuria aurantia*) Not only bacteria, fungi can also have phosphate solubilizing activities. In biofertilizers production and supply, NAU, Navsari has succeeded in winning the confidence of farmers by overcoming all the technical and administrative hurdles within a period of short span. Government of Gujarat introduced a novel approach, Krishi Mahotsava to connect the scientists with the farmers. This approach expedites the process of technology diffusion in the different parts of the state.

In an estimate, this biofertilizers have covered 66,600 ha land area and saved use of 295.7 tonnes urea. According to the criteria of IPCC, by saving 295.7 tonnes of urea, 59.14 tonnes CO2 emission has been reduced. Similarly, use of the phosphate solubilizing bacteria and potash mobilizing bacteria also have saved considerable CO_2 emission and helped in protection of environment.

Implications/Strategic Recommendations

The research work and extension activities carried out at NAU, Navsari clearly indicates that the biofertilizers viz., Nitrogen fixating (*Azotobacter chroococcum, Rhizobium* spp., *Gluconacetobacter diazotrophicus. Azospirillum brasilense, Azotobacter chroococcum*) Phosphate solubilizing bacteria (*Bacillus* spp. and *Pseudomonas* spp.) and Potash mobilizing bacteria (*Frateuria aurantia*) have high potential to save upto 50 per cent use of chemical fertilizers and increase crop yield by 20 per cent. These biofertilizers also have high potential to reduce the Green House Gases and protect the health of plant, animal, human and environment.

References*

1. Anzai, Y. Kim, H., Park, J. Y., Wakabayashi, H. and Oyaizu, H. 2000. Phylogenetic affiliation of the pseudomonad based on 16S rRNA sequence. *Int J Syst Evol Microbiol*.50: 1563-1589.
2. Chaudhary, A. 2023. Unleashing the Potential of Single Super Phosphate. Indian Journal of Fertilisers 19 (3): 178-179
3. Harpreet Kaur. 2019. Forms of Potassium in Soil and their Relationship with Soil Properties- A Review. *Int. J. Curr. Microbiol. App. Sci.* 8(10): 1580-1586. doi: https://doi.org/10.20546/ijcmas.2019.810.184

4. Jaeger, K.E., Ransac, S., Dijkstra, B. W., Colson, C., Heuvel, M. and Misset, O. 1994. Bacterial Lipases. *Fems Microbiol Rev.* 15: 29-63.
5. Khunt, M.D., Solanki V.A., Sabalpara A.N., and Mahatma L. 2014. Role of biofertilizers in plant nutrient management under present scenario in the book Major Constraints and Verdict of Crop productivity edited by UC Bhale. by *Astral International Pvt Ltd, New Delhi* ISBN No-978-93-5124-348-9.
6. Khunt M.D., Solanki V.A. and Mahatma L. 2015. Use of Biofertilizers in Organic Farming in the edited book Horticultural Technology Management edited by Udit Kumar and Birendra Prasad published by *Jaya Publishing House Publisher and Distributor* 27-B, Pocket-B, Dilshad Garden, Delhi-110095 (India).
7. Kim, G.W., Alam, M.A., Lee, J.J. Kim, G.Y., Kim, P.J. Khan, M.I. 2017. Assessment of direct carbon dioxide emission factor from urea fertilizer in temperate upland soil during warm and cold cropping season, *European Journal of Soil Biology*, Volume 83, Pages 76-83, https://doi.org/10.1016/j.ejsobi.2017.10.005.
8. Kojima, Y., Kobayashi, M. and Shimizu. S. 2003. A Novel Lipase from Pseudomonas fluorescens HU380: Gene Cloning, Overproduction, Reneturation-Activation, Two-Step Purification, and Characterization. *J biosci bioeng.* 96(3): 242-249.
9. Kumari, A., Naik, V. R., Patel, P. B. and Vala, J. R. 2025. Effect of Brown Manuring and Herbicide Application on Weed Parameters on Aerobic Rice under Varying Planting Geometry. *Journal of Advances in Biology & Biotechnology* 28 (1):1005-10. https://doi.org/10.9734/jabb/2025/v28i11957.
10. Mahatma L, Sharma J.K., Patel H.P., Patel N.M. and Patel R.P. 2023. Diversity of PGPM and ecosystem services. in the edited book Plant growth promoting microorganisms of arid region. edited by Mawar R., Sayyed R.Z., Sharma S.K., Sattiraju K.S. Springer Nature Singapore, Singapore, pp 93–124. https://doi.org/10.1007/978-981-19-4124-5_5
11. Mahatma L., Makwana K.V. and Sabalpara A.N. 2016a. Enhancement of sugarcane production and productivity by the biofertilizers with graded chemical fertilizers. *Indian Journal of Sugarcane Technology* 31(01): 6-9.
12. Mahatma L., Naik B.M., Mehta B.P., Solanky K.U., Chaudhary P.P. and Sabalpara A.N. 2016b. Field efficacy of different isolates of *Azotobacter croococum* for improving the yield of finger millet (*Eleusine coracana* (L.) Gaertn. *World Journal of Pharmaceutical and Life Sciences Vol. 2, Issue 5, 285-291.*
13. Mahatma, L., Khunt, M.D. Patel, R.P. and Mahatma, M.K. 2019. Legume-Rhizobium Symbiosis: Partially Explored and Exploited Interaction in book Plant Growth Promoting Microorganisms edited by Niranjan S. R. and Udayashankar, A. C. Nova Publisher pp 515-27.
14. Panchal B.J., Patel, S., Rajkumar, Jha S., Mahatma L. and Singh D. 2015. Isolation and identification of phosphate solubilizing *Penicillium expansum* NAUG-B1 and their consequence on growth of Brinjal. *Ecology, Environment and Conservation* 21 (December Suppl.) pp. S259-S267.
15. Sabalpara A. N. and Mahatma L. 2016. Effective Utilization of Biofertilizers in Horticultural Crops in the edited book Commercial Horticulture edited by N.L. Patel, S.L. Chawla and T.R. Ahlawat *New India Publishing Agency, New Delhi, India.*
16. Sabalpara A.N. and Mahatma L. 2017. Role of microbes in floriculture Book chapter in the book Advances in Floriculture and Landscape Gardening edited by Suresh K Malhotra and Lallan Ram. Published by *Government of India, Medziphema, Dimapur, Nagaland* – 797106 pp 205-211.
17. Sabalpara, A.N. and Mahatma L. 2019. Role of Microbes in Sustainable Agriculture Natural Resource Management: *Ecological Perspectives*, 147-163.

18. Topiwala M. and Mahatma L. 2017. Molecular Characterization of Oil Degrading Bacteria *Trends in Biosciences,* 10(2), 859-862.
19. Zayed O, Hewedy O.A., Abdelmoteleb A., Ali M., Youssef M.S., Roumia A.F., Seymour D., Yuan Z.C. 2023. Nitrogen Journey in Plants: From Uptake to Metabolism, Stress Response, and Microbe Interaction. *Biomolecules.* Sep 25;13(10):1443. doi: 10.3390/biom13101443. PMID: 37892125; PMCID: PMC10605003.

* The content of this chapter was prepared in consultation with above references.

7

Successful Woman Entrepreneur in Mushroom Cultivation in Tapi District (Gujarat)

H.R. Jadav[1], K.N. Rana[2], A.J. Dhodia[3], C.D. Pandya[4] & M.D. Lad[5]

[1]Scientist (Plant Protection), [2]Scientist (Crop Production), [3]Scientist (Agricultural Extension), [4]Senior Scientist & Head, [5]Assistant Extension Educationist, Directorate of Extension Education, Krishi Vigyan Kendra, Navsari Agricultural University

Email: hrjadav@nau.in

Abstract

Smt. Anjanaben Gamit is a Civil Engineer-Turned-Women Entrepreneur in Mushroom Cultivation. Smt. Anjanaben Gamit used to lead a normal life in the society. But, later on, she was successful in realizing her dream to secure livelihood in general and tribals, in particular marginal land. An article on Oyster Mushroom Cultivation published by the Krishi Vigyan Kendra, Tapi in the Agro-Sandesh gave direction to her dreams. Following this, she visited the KVK-Tapi and opted for the Mushroom cultivation under the guidance of KVK scientists. Moving forward to realize her dream, she joined a four-day vocational training program on "Entrepreneurship development through Mushroom Cultivation" at the KVK, Tapi and decided to initiate the mushroom cultivation during 2017 at available resources with the technical guidance from KVK-Tapi. For this, she prepared a mushroom growing house in the parking shed by using bamboo and green shade net along four sides. She was also supplied with all the inputs, viz., spawn (mushroom seed), polythene bags, seeds and chemicals (Carbendazim and formalin) along with the follow-up visits and technical guidance by the KVK scientists. On starting the mushroom cultivation for the first time in October, 2017, she harvested about 140 kg Mushroom with a value of Rs. 28,000/- in a simple small low cost shed (Size 15' x 10') within 2.5 months by investing Rs. 11,000 as production cost.

Smt. Anjanaben's success in mushroom production from October, 2017 to March, 2019 and 18 months' experience in mushroom cultivation motivated her to extend the mushroom production unit. So, she enlarged his mushroom house (size of 23'x80') by investing additional Rs. 1,72,000/- during 2019-20. With all the support from the relatives, social contacts and based on demand, she packed 100 to 200 grams' packets and sold them in Vyara town through Anganwadi workers, retail shopkeepers and vegetable vendors. The telephonic booking of mushrooms, made the marketing easier. She also started the sale of mushrooms from the "Organic market desk - selling organic produce from organic producer to direct consumer" commenced by Collector, Tapi District. For her achievements, Smt. Anjanaben not only got recognition in her nearby areas, but has also been felicitated by the Government of India too. From April, 2019 to December, 2019, she used 250 kg spawn and produced 1,234 kg of mushroom with a gross income of Rs. 3,08,500/-. The total cost of production was Rs. 88,350/-. By this way, she earned a net profit of Rs. 2,20,150/- during 2019-20. Now onward, she used 1340 kg spawn and produced 3700 kg of fresh mushroom with a gross income of Rs. 7,36,500/- from April, 2021 to February, 2025. The total cost of production was Rs. 3,94,650/-. By this way, she earned a net profit of Rs. 3,62,400/- during 2021-25. She was started oyster mushroom spawn production in her laboratory since 2022 and she was produced 300 kg spawn per month and sell them nearby local area at Rs.100 per one kg. She got net profit Rs. 3,57,900 from selling 11930 kg mushroom spawn since last three years. She sells 300 kg dried mushroom powder in market at price Rs. 1500 per kg during the year 2022-23, 2023-24 and 2024-25. Thus she got Rs. 4,29,000 by selling the dried mushroom powder.

Keywords: *Oyster Mushroom cultivation, Women Entrepreneur*

Introduction

The Pleurotus species, belongs of the basidiomycetes class, are typically grown on non-composted lignocellulose substrates and have a shorter growth cycle compared to other types of mushrooms (Bellettini *et al.,* 2018). These are shell-like fruiting bodies commonly known as "oyster mushroom" and also referred as "Dhingri mushroom" in India. The fanor spatula-shaped fruit body can be found in various colours depending on the species, including white, pink, yellow, grey, and blue (Verma *et al.,* 2023). Their cultivation is popular due to the simple and low-cost methods. These mushrooms are highly favored not only for their taste, flavor, and pleasing texture but also for their nutritional and medicinal benefits (Sharma and Khanna 2019, Sarita *et al.,* 2023). Pleurotus sp. are found naturally in both temperate and tropical regions across the globe which are cultivated more widely around the world due to their adaptable nature (Badshah *et al.,* 2021). Despite their potential, India's mushroom production has declined significantly, from 487 million MT in 2017 to 314.84 million MT in 2023 (Anonymous, 2023). Mushroom farming in Gujarat presents several opportunities due to various factors such as favorable climatic conditions, easy availability of raw material increasing market demand and supportive government initiatives. In India, rice straw is mainly used as substrate for large scale mushroom production (Kundu *et al.*, 2022). But, cultivation and consumption of mushroom were limited due to lack of awareness about the health benefits of this high valued food. However, by the efforts of KVKs under Agricultural Universities through the trainings and awareness programs, mushroom cultivation becomes popular among the people of Gujarat particularly in rural areas. Entrepreneurs and farmers can tap into this growing sector for profitable ventures, provided they are equipped with the necessary knowledge and resources.

About the Technology Addressed

Smt. Anjanaben is the first rural farm women who can successfully produce mushroom in South Gujarat especially in Tapi district. Mushroom plays a major role of nutritious component of rural and urban people's daily diet. Paddy is the major food crop in India and also in Gujarat and paddy straw is also available as by-products in large quantity. Farmers generally used this paddy straw as animal feed. Farmers will get additional income with agriculture and animal husbandry, if they this paddy straw for mushroom cultivation. There is also good demand in market for mushroom and south Gujarat climate is also favorable for mushroom cultivation. But there are very few mushroom growers. As per the opinion of Scientists, South Gujarat is the very good zone in India for mushroom cultivation. Paddy straw or wheat straw and mushroom spawn are the major requirements in mushroom cultivation and if mushroom grower can take appropriate care during cultivation, they will earn maximum profit with satisfaction. By considering all this, Anjanaben adopted mushroom cultivation.

Anjanaben is extremely talented, hard worker & skilled lady. It was one dream of Mrs. Anjanaben Gamit to do something without land/marginal land for securing livelihood in general and tribals particular. She wanted to be independent and carve out and identify for herself. Meanwhile, she read an article on Oyster Mushroom cultivation published by Krishi Vigyan Kendra-Tapi in Agro-Sandesh dated 20th February, 2017. Then she visited KVK and decided to go for Mushroom cultivation under the guidance of KVK Tapi and joining with KVK, proved to be a boon for her.

Extension Management Practices

Initiatives

Smt. Anjanaben joined four days (19.09.2017 to 22.09.2017) training program on "Entrepreneurship development through Mushroom Cultivation" at KVK, Tapi and decided to initiate the mushroom cultivation during 2017 at available resources with the technical guidance from KVK, Vyara. During 2018-19, she also joined 200 hrs (27 days-01/12/2018 to 26/12/2018) training program of '**Mushroom Grower**" at KVK, Tapi sponsored by Agricultural Skill Council of India (ASCI) under RKVY to update her knowledge. Consequently, near his home there was a parking shed and she had decided to grow mushroom production at small scale in this parking shed. KVK, Vyara has been supported her for paddy straw cutting in chaff cutter demonstration unit. She prepared mushroom growing house in his parking shed by using bamboo and green shade net along four sides. All the inputs *viz*., spawn (mushroom seed), polythene bags, seeds and chemicals (Carbendazim & formalin) has been supplied by KVK, Tapi. Follow up visit, diagnostic visit has also been made by Scientists of KVK-Tapi. Regular guidance on telephone has also been provided

Reach

Krishi Vigyan Kendra, Vyara disseminated mushroom cultivation technology through various extension activities in different villages of Tapi district. KVK, also provided all the inputs required for mushroom cultivation. All the technical supports through trainings, field visits, diagnostic visits, telephone helpline were provided by KVK, Tapi. The 'Mushroom Grower" 200 hrs training program sponsored by Agricultural Skill Council of India (ASCI) under RKVY were also helpful for Anjanaben to update her knowledge. Moreover, "Organic market desk- Selling organic produce from organic producer to direct consumer" commenced by Collector of Tapi district in collaboration with DAO, ATMA and DHO Tapi were also proved to be good marketing source for her.

Output

During 2017-18, she started mushroom cultivation first time in October 2017 and harvest about 140 kg Mushroom with a value of Rs.28000/- in a simple small low cost shed (Size 15' x 10') within 2.5 months by investing Rs.11000 as production cost (Table.1) This encouraging results motivated her to start mushroom cultivation on regular basis and design a structure (Size 30' x 20') based on one-time expenditure of Rs.1,00,000/-(Table 2). From April 2018 to March 2019, she harvested total 1079 kg of mushroom with a value of Rs.3,77,650/- (Table 3). She got prevailing market price of Rs. 350/- per kg. She was benefitted in a good way. By this way, she proved her dream true with the adoption of mushroom cultivation and enough to convey a message to secure livelihood in tribal areas without land /marginal land. Anjanaben's success in mushroom production from October 2017 to March 2019 and 18 months' experience in mushroom cultivation motivate herself to extend mushroom production unit. So, she enlarged his mushroom house (size of 23'x80') by investing additional Rs. 172000/- during 2019-20 (Table. 2). From

April-2019 to December 2019, she used 250 kg spawn and produced 1234 kg of mushroom with a gross income of Rs. 3,08,500/-. Total cost of production was Rs. 88,350/-. By this way she gets net profit of Rs. 2,20,150/- during 2019-20 (Table 4).

During the Corona time, she meets to KVK scientists and showed her interest in oyster mushroom spawn production training. She wants to take benefit this period and joined training on mushroom spawn production at CoA, NAU, Waghai during 2019. After that, she spends Rs. 11.0 lakhs for establishment of 'Mushroom Spawn Production Laboratory'. Aside from training and other forms of support, the Gujarat government has also been providing a 75% subsidy on Mushroom Spawn Production Laboratory. Now she was started oyster mushroom spawn production in her house by getting govt. subsidies about 8.5 lakhs. Initially, she started making mushroom spawns with minimal investment. She purchased a small autoclave and used a spirit lamp as a substitute for laminar to keep costs low. With this simple setup, she was able to produce 10 kg of spawn every day. With new skills and more experience, she expanded her mushroom operations. Now, she was produced 300 kg spawn per month and sell them nearby local area at Rs.100 per one kg. She got net profit Rs. 3,57,900 from selling 11930 kg mushroom spawn since last three years.

From April-2020 to December 2020, she used 250 kg spawn and produced 1234 kg of mushroom with a gross income of Rs.3,08,500/-. Total cost of production was Rs. 88,350/-. By this way she gets net profit of Rs.2,20,150/- during 2020-21 (Table.5). From April-2021 to December 2021, she used 280 kg spawn and produced 1400 kg of mushroom with a gross income of Rs.2,24,000/-. Total cost of production was Rs. 91,950/-. By this way she gets net profit of Rs.1,32,050/- during 2021-22 (Table.6). From April-2022 to December 2022, she used 450 kg spawn and produced 2250 kg of mushroom with a gross income of Rs. 2,50,000/-. Total cost of production was Rs. 1,12,500/-. By this way she gets net profit of Rs. 1,58,050/- during 2022-23 (Table.7). From April-2023 to December 2023, she used 310 kg spawn and produced 1550 kg of mushroom with a gross income of Rs. 1,37,500/-. Total cost of production was Rs. 95700/-. By this way she gets net profit of Rs. 41,800/- during 2023-24 (Table.8). From April-2024 to February 2025, she used 300 kg spawn and produced 1500 kg of mushroom with a gross income of Rs. 1,25,000/-. Total cost of production was Rs. 94500/-. By this way she gets net profit of Rs. 30,500 /- during 2023-24 (Table.8).

Now, she was constructed pakka structure (6000 sq. ft) with cost Rs. 11.0 lakhs for mushroom production as well as training project at her home during 2024-25. Gujarat government has also been providing a 75% subsidy on Mushroom about 4.6 lakhs for her mushroom pakka shed house. She dried 1000 kg fresh mushroom under sunlight since last three years and used them for powder processing. She sells 300 kg dried mushroom powder in market at price Rs. 1500 per kg during the year 2022-23, 2023-24 and 2024-25. Thus she got Rs. 4,29,000 by selling the dried mushroom powder. The monthly details of cost of cultivation, mushroom production and profit earn during 2017-18, 2018-19, 2019-20, 2020-21, 2021-22, 2022-23, 2023-24 and 2024-25 were mentioned in Annexure-1, Annexure-2, Annexure-3, Annexure-4, Annexure-5, Annexure-6, Annexure-7 and Annexure-8, respectively.

Outcome/Impact

Soon, Anjanaben began to see substantial progress; her efforts began to yield bountiful harvests. As her mushrooms grew, and her farming methods improved, the next challenge she had to tackle was marketing. With the help of relatives, social contacts and based on demand she packed 100 to 200-gram packet and sold in Vyara town through Anganwadi worker, retail shopkeepers, and vegetable vendors. Consumers also have been booked mushroom on telephone. That approach made marketing very easy for her. She also started mushroom selling from the "Organic market desk- selling organic produce from organic producer to direct consumer" commenced by Hon. Collector, Tapi district. Now she is a successful entrepreneur and became a role model to other women in the village as well as other village. She became a master trainer for all the farm women in Vyara block. For her achievements, Smt. Anjanaben not only got recognition in her nearby areas, but has also been felicitated by the Government of India too.

Award/Recognitions Received

i. She also inspired by Hon.VC and Hon.DEE of NAU Navsari. She also shared her experience in "Innovative Farmers Meet Programme" of KVK, Vyara and KVK, Navsari during 18th June, 2018 at KVK, Tapi

ii. Due to outstanding contribution in Mushroom cultivation, Mrs. Anjanaben also honored by Smt. Smriti Irani, Hon. Union Textile Minister, Govt. of India at KVK, Tapi during 21st Sept., 2018.

iii. Her success story was also telecasted in DD Girnar news channel and in Dhartiputra program of TV-9 Gujarati channel during 11th February, 2019 and 13th March, 2019, respectively.

iv. She also shared his experience of mushroom cultivation in" Khedut Mela" program jointly organized by ATMA and FTC Tapi during 12th January, 2020. Mrs. Anjanaben also given award by Plant Protection Association of Gujarat during 2024

Mentoring Fellow Farmers

As her business flourished in the next couple of years, the entrepreneur realised the importance of sharing her knowledge with others. In 2022, Anjanaben began offering mushroom farming training to aspiring individuals in her village. With the support of the DDO-Tapi- Department, she was able to establish and formalize these classes. she said that it felt good to teach them the process and see them start their own businesses. Participants in the training program also purchased spawns from me at Rs 100 per kg. She charged Rs 1000/- per person for the training sessions and 15-20 persons was trained every two months

Don't be afraid to take that first step

Now, Anjanaben's business generates a steady income through multiple streams. She earns around Rs 10,000-15,000 monthly from training sessions. Her mushroom spawn sales contribute approximately Rs 30,000. Additionally, she sells 15-20 kg of fresh mushrooms daily at Rs 150 per kg, generating Rs 15,000-20,000 each month. She also helps fellow farmers by buying their mushrooms at a slightly lower price and reselling them, creating a mutually beneficial arrangement. All of these efforts bring her a monthly turnover of around Rs 2 lakh. In addition to her core business, Anjanaben has introduced value-added product i.e., dried mushroom powder which has been well received by customers. She also encourages fellow farmers in her village to set up mushroom cultivation, providing them with an opportunity to boost their sales and diversify their income

Conclusion

Now, Anjanaben is a successful mushroom grower with expertise in mushroom cultivation. She is a well known progressive mushroom grower for the farmers of nearby villages and rural youths. By witnessing her success, farmers of nearby villages, rural youths, students, scientists and also the dignitaries visited her mushroom unit. She inspired and facilitated many farmers to start mushroom farming. Five to six unwaged youths also get employed through Anjanaben's mushroom production unit. Maximizing yields through optimized production processes can significantly boost profits. Establishing sales channels that connect directly with consumers can enhance profit margins by reducing distribution costs. Moreover, implementing sustainable mushroom farming practices not only appeals to health-conscious consumers but can also reduce operational costs, leading to increased profitability. For example, utilizing waste materials for substrate can lower production costs while promoting eco-friendly practices. Conduct regular market research to stay updated on consumer preferences and emerging trends in the mushroom industry. Invest in technology for data analytics to monitor growth conditions and improve operational efficiency. Explore partnerships with local restaurants to create dedicated supply channels that enhance your visibility and sales. In conclusion, the profit potential of a mushroom cultivation business is substantial, provided that effective strategies are implemented to overcome challenges and capitalize on market opportunities.

Implications and Strategic Recommendations

Women play a significant role in socio-economic development of family and society. They are also recognized as an integral part of agriculture and allied activities. As an entrepreneur they have a major role in the growth of their enterprise. Mushroom farming using agri-waste as substrates can offer a sustainable solution to the food security challenges of inadequate and imbalanced diets. Developing strategies to exploit the potential of the mushroom industry fully is yet to be explored in Gujarat. Mushrooms are used not only as a source of nutrients, but also as medicinal resources. Polysaccharides from mushrooms were reported to exhibit immunomodulation properties, antitumour, antioxidant, antimicrobial and prebiotic activity. The benefits of mushrooms are relatively economical because the mushrooms can be grown on a number of inexpensive agricultural wastes such as rice straw, corn cobs and saw dust. In the quest for economical and ecologically sound methods for environmental remediation, the use of mushrooms is a very good approach and solution. Management from 'waste-to-wealth' is essential for more sustainable farming globally, and increasing mushroom production in India seems a viable and attractive option. Boosting the commercial value of products whether in a fresh or processed form could increase concentration of demand and encourage market orientation.

Table 1. Details of expenditure and income from mushroom cultivation (2017-18)

S. No.	Cost of Cultivation (Rs.)		Total Production (Kg)	Gross Income (Rs.)	Net Income (Rs.)
A	**Non-recurring expenditure**				
1	Mushroom house (bamboo for making racks, gunny bags)	2,500	140 kg	28000 (Rs.200/kg)	17000
2	Spray pump	2,000			
3	Other miscellaneous exp. (tubs and drums)	600			
	Total	**5,100**			

B	**Recurring expenditure (cost of raw material)**				
1	Paddy straw	800			
2	Sugarcane Bagas	300			
3	Polythene bags (Rs.4 per bag)	420			
4	Spawn 24 kg (Rs.120/- per kg)	2,880			
5	Formaline 5.2 lit (Rs.100/- per lit)	520			
6	Carbendazim 0.3 kg(Rs.600/- per kg)	180			
7	Other	800			
8	**Total**	**5,900**			
	Total Production cost (A+B)	**11,000**			

Table 2. Details of expenditure for preparation of well-designed pucca mushroom house (2018-19 and 2019-20)

S. No.	Particulars	Cost of Cultivation (Rs.)		
		2018-19	**2019-20**	**Total**
1	Mushroom house	40,000 (20'x30')	1,30,000 (23'x80')	1,70,000
2	Bamboos for preparation of side wall and racks	22,000	13,000	35,000
3	Spray pumps, gunny bags	7,000	3,000	10,000
4	Sprinkler/Fogger system	15,000	16,000	35,000
5	Irrigation Tank with motor	7,000	8,000	15,000
6	Tubs and drums and other miscellaneous	9,000	2,000	7,000
	Total (A)	1,00,000	1,72,000	2,72,000

Table 3. Expenditure for upgrading mushroom house and economics of mushroom cultivation (2018-19)

S. No.	Particulars with Cost of Cultivation (Rs.)		Total Production (Kg.)	Gross Income (Rs.)	Net Income (Rs.)
1	Mushroom house	1,00,000	1,079 kg	3,77,650.00 (Selling price @ 350/- per kg)	2,11,495.00
2	Paddy straw 6400 no.@ Rs.3/-per no.	19,200			
3	Sugarcane Bagas 2.0 ton	5,000			
4	Polythene bags 1350 no@ Rs.2 per bag	2,700			
5	Spawn (260 kg @ 120/- per kg)	31,200			
6	Formaline 50 lit @ 100/- per lit	5,000			
7	Carbendazim 4.0 kg @ 600/- per kg	2,400			
8	other miscellaneous exp.	1,000			
	Total	1,66,500			

Table 4. Expenditure and economics of mushroom cultivation (2019-20)

S. No.	Particulars with Cost of Cultivation (Rs.)		Total Production	Gross Income (Rs.)	Net Income (Rs.)
1	Paddy straw 6000 nos @ Rs.3/-per no.	18,000	1,234 kg	3,08,500.00 (Selling price @ Rs. 250/- per kg)	2,20,150.00
2	Sugarcane Bagas 2.0 ton	5,000			
3	Wheat straw 1500 kg	1,500			
4	Polythene bags 1000 no.@ Rs.2 per bag	2,000			
5	Spawn 250 kg @ 120/- per kg	30,000			
6	Formaline 40 lit @ 100/- per lit	4,000			
7	Carbendazim 4.0 kg @ 600/- per kg	2,400			
8	Labour charges and other	25,450			
	Total (B)	88,350			

Table 5. Expenditure and economics of mushroom cultivation (2020-21)

S. No.	Particulars with Cost of Cultivation (Rs.)		Total Production	Gross Income (Rs.)	Net Income (Rs.)
1	Paddy straw 6000 nos @ Rs.3/-per no.	18,000	1,234 kg	3,08,500.00 (Selling price @ Rs. 250/- per kg)	2,20,150.00
2	Sugarcane Bagas 2.0 ton	5,000			
3	Wheat straw 1500 kg	1,500			
4	Polythene bags 1000 no.@Rs.2 per bag	2,000			
5	Spawn 250 kg @ 120/- per kg	30,000			
6	Formaline 40 lit @ 100/- per lit	4,000			
7	Carbendazim 4.0 kg @ 600/- per kg	2,400			
8	Labour charges and other	25,450			
	Total (C)	88,350			

Table 6. Expenditure and economics of mushroom cultivation (2021-22)

S. No.	Particulars with Cost of Cultivation (Rs.)		Total Production	Gross Income (Rs.)	Net Income (Rs.)
1	Paddy straw 6000 nos @ Rs.3/-per no.	18,000	1,400 kg	2,24,000.00 (Selling price @ Rs. 160/- per kg)	1,32,050.00
2	Sugarcane Bagas 2.0 ton	5,000			
3	Wheat straw 1500 kg	1,500			
4	Polythene bags 1000 no.@Rs.2 per bag	2,000			
5	Spawn 280 kg @ 120/- per kg	33,600			
6	Formaline 40 lit @ 100/- per lit	4,000			
7	Carbendazim 4.0 kg @ 600/- per kg	2,400			
8	Labour charges and other	25,450			
	Total (D)	91,950			

Table 7. Expenditure and economics of mushroom cultivation (2022-23)

S. No.	Particulars with Cost of Cultivation (Rs.)		Total Production	Gross Income (Rs.)	Net Income (Rs.)
1	Paddy straw 6000 nos @ Rs.3/-per no.	18,000	2,250 kg	2,50,000.00 (Selling price @ Rs. 200/- per kg)	1,58,050.00
2	Sugarcane Bagas 2.0 ton	5,000			
3	Wheat straw 1500 kg	1,500			
4	Polythene bags 1000 no.@Rs.2 per bag	2,000			
5	Spawn 450 kg @ 120/- per kg	54,000			
6	Formaline 40 lit @ 100/- per lit	4,000			
7	Carbcndazim 4.0 kg @ 600/- per kg	2,400			
8	Labour charges and other	25,600			
	Total (E)	1,12,500			

Table 8. Expenditure and economics of mushroom cultivation (2023-24)

S. No.	Particulars with Cost of Cultivation (Rs.)		Total Production	Gross Income (Rs.)	Net Income (Rs.)
1	Paddy straw 6000 nos @ Rs.3/-per no.	18,000	1,550 kg	1,37,500.00 (Selling price @ Rs. 250/- per kg)	41,800.00
2	Sugarcane Bagas 2.0 ton	5,000			
3	Wheat straw 1500 kg	1,500			
4	Polythene bags 1000 no.@ Rs.2 per bag	2,000			
5	Spawn 310 kg @ 120/- per kg	37,200			
6	Formaline 40 lit @ 100/- per lit	4,000			
7	Carbendazim 4.0 kg @ 600/- per kg	2,400			
8	Labour charges and other	25,600			
	Total (F)	95,700			

Table 9. Expenditure and economics of mushroom cultivation (2024-25)

S. No.	Particulars with Cost of Cultivation (Rs.)		Total Production	Gross Income (Rs.)	Net Income (Rs.)
1	Paddy straw 6000 nos @ Rs.3/-per no.	18,000	1,500 kg	1,25,000.00 (Selling price @ Rs. 250/- per kg)	30,500.00
2	Sugarcane Bagas 2.0 ton	5,000			
3	Wheat straw 1500 kg	1,500			
4	Polythene bags 1000 no.@ Rs.2 per bag	2,000			
5	Spawn 300 kg @ 120/- per kg	36,000			
6	Formaline 40 lit @ 100/- per lit	4,000			
7	Carbendazim 4.0 kg @ 600/- per kg	2,400			
8	Labour charges and other	25,600			
	Total (G)	94,500			

Table 10. Month-wise economics of mushroom cultivation (2017-18)

S. No.	Month	Production (Kg)	Expenditure (Rs.)	Gross income (Rs.)	Net income (Rs.)
1	October-17	50	3,500	10,000	6,500
2	November-17	40	3,500	8,000	4,500
3	December-17	50	4,000	10,000	6,000
	Total	**140**	**11,000**	**28,000**	**17,000**

Table 11. Month-wise economics of mushroom cultivation (2018-19)

S. No.	Month	Production (Kg)	Expenditure (Rs.)	Gross income (Rs.)	Net income (Rs.)
1	April-18	60	9,240	21,000	11,760
2	May-18	85	13,090	29,750	16,660
3	June-18	0	0	0	0
4	July-18	80	12,320	28,000	15,680
5	August-18	80	12,000	28,000	16,000
6	Seppember-18	90	13,860	31,500	17,640
7	October-18	130	20,075	45,500	25,425
8	November-18	150	23,100	52,500	29,400
9	December-18	165	25,410	57,750	32,340
10	January-19	127	19,560	44,450	24,890
11	February-19	82	12,700	28,700	16,000
12	March-19	30	5,145	10,500	5,355
	Total	**1079**	**1,66,500**	**3,77,650**	**2,11,150**

Table 12. Month-wise economics of mushroom cultivation (2019-20)

S. No.	Month	Spawn used(Kg)	Mushroom Production (Kg)	Expenditure (Rs.)	Gross income (Rs.)	Net income (Rs.)
1	April-19	23	91	8,050	22,750	14,700
2	May-19	30	123	10,500	30,750	20,250
3	June-19	50	225	17,500	56,250	38,750
4	July-19	0	0	0	0	0
5	August-19	0	0	0	0	0
6	September-19	0	0	0	0	0
7	October-19	40	200	14,000	50,000	36,000
8	November-19	70	350	24,500	87,500	63,000
9	December-19	35	245	13,800	61,250	47,450
	Total	**248**	**1234**	**88,350**	**3,08,500**	**2,20,150**

Table 13. Month-wise economics of mushroom cultivation (2020-21)

S. No.	Month	Spawn used (Kg)	Mushroom Production (Kg)	Expenditure (Rs.)	Gross income (Rs.)	Net income (Rs.)
1	April-20	30	150	10,500	22,500	14,700
2	May-20	30	150	10,500	30,750	20,250
3	June-20	30	150	10,500	22,500	38,750
4	July-20	0	0	0	0	0
5	August-20	0	0	0	0	0
6	September-20	0	0	0	0	0
7	October-20	50	250	14,000	37,500	36,000
8	November-20	50	250	24,500	37,500	63,000
9	December-20	30	150	13,800	22,500	47,450
	Total	**248**	**1234**	**88,350**	**3,08,500**	**2,20,150**

Table 14. Month-wise economics of mushroom cultivation (2021-22)

S. No.	Month	Spawn used (Kg)	Mushroom Production (Kg)	Expenditure (Rs.)	Gross income (Rs.)	Net income (Rs.)
1	April-21	50	250	16,223	40,000	23777
2	May-21	50	250	16,223	40,000	23,777
3	June-21	50	250	16,223	40,000	23,777
4	July-21	0	0	0	0	0
5	August-21	0	0	0	0	0
6	September-21	0	0	0	0	0

7	October-21	50	250	16,223	40000	23,777
8	November-21	50	250	162,23	40,000	23,777
9	December-21	30	150	10,835	24,000	13,165
	Total	**280**	**1400**	**91,950**	**2,24,000**	**1,32,050**

Table 15. Month-wise economics of mushroom cultivation (2022-23)

S. No.	Month	Spawn used (Kg)	Mushroom Production (Kg)	Expenditure (Rs.)	Gross income (Rs.)	Net income (Rs.)
1	April-22	60	300	13,500	48,000	34,500
2	May-22	60	300	13,500	48,000	34,500
3	June-22	60	300	13,500	48,000	34,500
4	July-22	30	150	10,500	24,000	13,500
5	August-22	30	150	10,500	24,000	13,500
6	September-22	30	150	10,500	24,000	13,500
7	October-22	60	300	13,500	48,000	34,500
8	November-22	60	300	13,500	48,000	34,500
9	December-22	60	300	13,500	48,000	34,500
	Total	**450**	**2250**	**1,12,500**	**3,60,000**	**2,47,500**

Table 16. Month-wise economics of mushroom cultivation (2023-24)

S. No.	Month	Spawn used (Kg)	Mushroom Production (Kg)	Expenditure (Rs.)	Gross income (Rs.)	Net income (Rs.)
1	April -23	60	300	13,500	48,000	34,500
2	May -23	60	300	13,500	48,000	34,500
3	June -23	60	300	13,500	48,000	34,500
4	July-23	0	0	0	0	0
5	August-23	0	0	0	0	0
6	September-23	0	0	0	0	0
7	October -23	50	250	16,900	40,000	23,100
8	November-23	50	250	24,500	40,000	15,500
9	December -23	30	150	13,800	24,000	10,200
	Total	**310**	**1550**	**95,700**	**2,48,000**	**1,52,300**

Table 17. Month-wise economics of mushroom cultivation (2024-25)

S. No.	Month	Spawn used (Kg)	Mushroom Production (Kg)	Expenditure (Rs.)	Gross income (Rs.)	Net income (Rs.)
1	April-24	60	300	16,800	51,000	34,200
2	May-24	60	300	16,800	51,000	34,200
3	June-24	60	300	16,800	51,000	34,200
4	July-24	60	300	16,800	51,000	34,200
5	August-24	0	0	0	0	0
6	September-24	0	0	0	0	0
7	October-24	0	0	0	0	0
8	November-24	0	0	0	0	0
9	December-24	0	0	0	0	0
10	January-25	30	150	13,650	25,500	11,850
11	February-25	30	150	13,650	25,500	11,850
	Total	**300**	**1500**	**94,500**	**2,55,000**	**1,60,500**

Fig 1. Mrs. Anjanaben working in her new pakka mushroom shed house (6000 Sq. ft.) constructed during 2024-25 at her village: Nani chikhali, Ta: Vyara, Dt: Tapi

Fig 2. KVK scientist provided regular guidance through frequently follow up visit, diagnosis field visit and telephonic advisory

Fig 3. Dr. S.K. Roy, Director, ATARI, Pune has appreciated Mrs. Anjanaben's work during his visit

Fig 4. Mrs. Anjanaben receiving award given by Plant Protection Association of Gujarat during 2024

Fig 5. Mrs. Anjanaben given guidance on mushroom cultivation in her various training sessions

Fig 6. Mrs. Anjanaben joined four days (19/09/2017 to 22/09/2017) training program on "Entrepreneurship development through mushroom cultivation" at KVK-Tapi

Fig 7. Smt. Smriti Irani, Hon. Union Textile Minister, Govt. of India visited to mushroom stall installed by Mrs. Anjanaben at KVK-Tapi on 21st Sept. 2018

Fig 8. Due to outstanding contribution in mushroom cultivation, Mrs. Anjanaben also honored by Smt. Smriti Irani, Hon. Union Textile Minister, Govt. of India at KVK-Tapi on 21st Sept. 2018

Fig 9. Mrs. Anjanaben initiated mushroom cultivation in her old house (size of 23'x80') during 2019-20 at her village: Nani chikhali, Ta: Vyara, Dt

Fig 10. Farmers of nearby villages, rural youths, students, scientists and also the dignitaries visited her mushroom cultivation house at village: Nani chikhali, Ta: Vyara

Fig 11. Mrs. Anjanaben started mushroom selling from the "Organic market desk-selling organic produce from organic producer to direct consumer" commenced by Hon. Collector, Tapi district

Fig 12. Mrs. Anjanaben also honored by Dr. G.R. Patel, Hon. DEE of NAU, Navsari during SAC meeting at KVK-Tapi on 31st January, 2020

Fig 13. Mrs. Anjanaben also inspired by Hon. VC and Hon. DEE of NAU, Navsari. She also shares her experience in "Innovative Farmers Meet Programme" of KVK, Vyara and KVK, Navsari on 18th June, 2018 at KVK-Tapi

Fig 14. Mrs. Anjanaben shared her experience of mushroom cultivation in Khedut Mela jointly organized by ATMA and FTC Tapi on 12th January, 2020

References*

1. Anonymous. 2023. Annual Report-2023, Directorate of Mushroom Research https://dmrsolan.icar.gov.in/AnnualReportDMR2023.pdf
2. Badshah, K., Ullah, F., Ahmad, B., Ahmad, S., Alam, S., Ullah, M. and Sardar, S. 2021. Management of Lycoriella ingenua (Diptera: Sciaridae) on oyster mushroom (Pleurotus ostreatus) through different botanicals. International Journal of Tropical Insect Science, 41, 1435-1440.
3. Bellettini, M. B., Bellettini, S., Fiorda, F. A., Pedro, A. C., Bach, F., Fabela-Morón, M. F. and Hoffmann-Ribani, R. 2018. Diseases and pests noxious to Pleurotus spp. mushroom crops. Revista Argentina de microbiologia, 50(2), 216-226

4 Kundu, K., Chandra, B., Viswavidyalaya, K., Mukhopadhyay, S., Chandra, B., Viswavidyalaya, K., ... & Dutta, S. 2022. Influence of different growth substrates and their combination on nutritional and mineral contents of Oyster mushroom (Pleurotus florida). Biological Forum – An International Journal, 14(4), 944-951

5. Sarita, Roop Singh, Kalpna Yadav, Karan Singh, Priyanka and Gagandeep Singh. 2023. Cultivation of Oyster Mushroom: A Review. Biological Forum – An International Journal, 15(8), 412-418.

6. Sharma, N. D. and Khanna, A. S. 2019. Feeding losses due to damaging potential of grubs and adults of Cyllodes indicus in Pleurotus sajor caju. Journal of Pharmacognosy and Phytochemistry, 8(4), 3178-3182

7. Verma, P., Nath, M., Sharma, A., Barh, A., Kamal, S., Sharma, V. P., and Kumar, A. 2023. Comparative evaluation of different spawn substrates on the growth and yield of oyster mushroom. Mushroom Research, 32(1), 41-49.

* The content of this chapter was prepared in consultation with above references.

8

Nauroji Stone House Fruit Fly Trap A Success Story

Z.P. Patel[1] , S.R. Patel[2] and J.J. Pastagia[3]

[1]Vice Chancellor, [2]Assistant Professor and [3]Professor and Head, Department of Entomology, N. M. College of Agriulture, Navsari Agricultural University, Navsari
Email: vc@nau.in

Abstract

South Gujarat is leading in fruits and vegetable production in Gujarat state. The incidence of fruit fly in fruits and cucurbit vegetable crops not only reduces the yield and quality but also causes considerable economic loss. Adults of the pest occur throughout the year. However, monitoring and management of immature stages of fruit fly in field is difficult as the maggots remain inside the fruit and pupation and overwintering occur in soil (Bansode and Patel, 2020). These flies are good fliers and thereby their spread is also very high and hence area wide adoption of management strategy can only be useful against these flies. Navsari Agricultural University has designed and commercialized an eco-friendly, economical and easily adoptable fruit fly trap popularly known as "Nauroji-Stonehouse Fruit Fly Trap" enables monitoring and mass trapping of male fruit flies through Male Annihilation Technique (MAT) giving better management in fruits and cucurbit vegetable crops for quality production.

Keywords: *Bactrocera spp., Fruit fly trap, Male Annihilation*

Introduction

Three districts of South Gujarat i.e., Surat, Navsari and Valsad are front-runners in horticulture cultivation. These are also part of an agri-export zone for fruits and vegetables. But, the life of these fruit- farmers was made miserable by tiny fruit fly which is a major pest of mango, sapota and cucurbits. Intensity of damage recorded was as high as over 30 per cent in mango and sapota, and between 20-40 per cent in cucurbitaceous vegetables. Fruit flies cause heavy damage in terms of quality and quantity to the farmers every year. Insecticidal sprays, control infestation little bit, but they are not only uneconomical but they leave residual effect much above tolerance limit, adversely affecting the prospects of export of such fruits and vegetables.

Fruit flies are a modern-day pestilence which ravage production. These flies, moreover, are good fliers and therefore their spread is extensive. Hence, area wide adoption of management strategy is required. Male Annihilation Technique (MAT), an Integrated Pest Management Technique (IPM), which uses sexual lures to capture males of fruit flies and kill them, is the only way to control growth and spread of fruit flies. This however, requires a community approach for implementation.

Navsari Agricultural University designed, developed and commercialized an eco-friendly, economical and easily adoptable fruit fly trap popularly known as "Nauroji-Stonehouse Fruit Fly Trap". The trap uses Methyl Eugenol and Cue lure as par pheromones well-known for fruit flies. In this trap, plywood blocks are soaked in the solution of lures + solvent + insecticide. Such traps remain effective for 5 to 6 months, which cover the entire fruiting period in mangos, sapota as well as cucurbitaceous vegetables and no recharge is necessary. It is eco- friendly, economical and easily adoptable.

It was decided to adopt this technology and spread it in the Surat- Navsari-Valsad belt for controlling and managing the fruit fly problem. It was also recognized that wide area adoption would help in developing this zone as a Pest Free Area (PFA). The project was taken up under RKVY.

About the Technology Addressed and Extension Management Practices

Nauroji-Stonehouse Fruits Fly Traps are prepared in food quality Testing Laboratory, NAU Navsari. Six talukas viz, Gandevi, Chikhali, Valsad, Pardi, Dharampur and Kaprada of Navsari and Valsad districts were selected for the purpose. A block of orchards in each village in each Taluka was selected for first year implementation. Other blocks were selected during the following years.

Fig 1. Preparation of plywood blocks

In the selected blocks, farmers of all categories were covered. Traps were distributed to the farmers considering their area of plantation of mango, sapota and cucurbitaceous vegetables. Farmer meetings were organized at village level to educate them regarding biology of fruit fly, nature of damage, technology for its management, installation of traps, etc. They were educated by delivering lectures using LCD projector, display of flex banner, booklet and actual demonstration of the traps. To generate awareness of technology, 93 farmer's group training meets were scheduled. During group meetings scientific information of fruit flies emphasizing the life cycle of the pest, host range, damage symptoms, severity of damage and control methods to combat the pest in eco-friendly manner were covered. Gram Panchayats, Village level co-operative societies and milk collecting centers collaborated in organizing meetings, preparing lists of farmers, disbursement of traps, etc.

With training and awareness, Farmers themselves installed the traps in their fields. Fruit fly population was also recorded at fortnightly intervals during the fruiting period in randomly selected orchards. The per cent damage of fruits was worked out from orchards where traps were installed as well as uninstalled orchards. The level of damage in treated and untreated orchards was compared to know the actual impact of technology.

In all, 209 villages in 6 talukas of two districts were selected as target area during 2009 and 2010. The project was implemented in about 6,814 ha area comprising of 6,367 ha of fruit orchards and 447 ha of melon orchards of 15,339 farmers of all categories. Around 1,10,640 traps were distributed to the farmers in the targeted villages for mango, sapota and cucurbitaceous vegetables. Farmers were also educated on maintaining sanitation in the orchard. The fallen/damaged fruits in the orchards were collected by the farmers and buried in the pits with application of methyl parathion dust to minimize further multiplication of fruit flies.

About 6000 booklets were disbursed to the farmers while 500 flex banners were displayed in the target villages. Data on fruit fly catches and infestation were recorded regularly in the target area. The project involved an investment □ 7.86 crores funded through RKVY.

Fig 2. Nauroji Sonehouse Fruit fly Trap

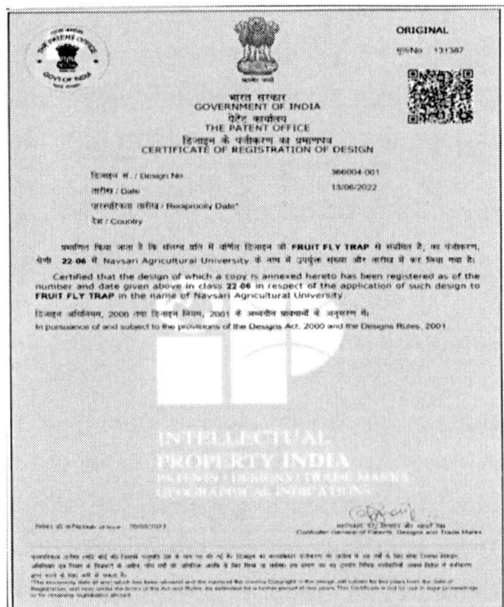

ORIGINAL

GOVERNMENT OF INDIA

THE PATENT OFFICE

CERTIFICATE OF REGISTRATION OF DESIGN

Design No 366004-001

Date 13/06/2022

Reciprocity Date*

Country

Certified that the design of which a copy is annexed hereto has been registered as of the number and date given above in class **22-06** in respect of the application of such design to **FRUIT FLY TRAP** in the name of Navsari Agricultural University.

In pursuance of and subject to the provisions of the Designs Act, 2000 and the Designs Rules, 2001

INTELLECTUAL PROPERTY INDIA

Fig 3. Design Patent Certifiate

By installing the fruit fly traps in a wide area, it was possible to successfully bring down the infestation level to 3 to 4 percent, which meant 85 per cent control of the pest. Using this technology, the fruit fly damage in mango orchards minimized to 3.06 (0 to 4) percent and the damage in untreated orchards was 30.34 (30 to 35%) per cent. Thus, more than 85.0 per cent damage due to fruit fly was reduced which resulted in increase of 27.27 per cent yield.

Moreover, being an eco-friendly approach, it was possible to certify the produce as organic. Thus, the quality of a large quantity of fruits and vegetables improved making them suitable for export. Low expenditure incurred on the control of fruit fly could motivate the growers and possibly enhance the sustainability of the technology demonstrated. The area wide adoption of fruit fly traps could effectively kill the male flies and thereby check further multiplication without disturbing the ecosystem. If such technology is adopted for a considerable time, it would be helpful in recognizing the area as PFA (Pest Free Area) and ultimately it will help boosting the export trade. About 15,339 farmers of all categories as well as free riders of six talukas in two districts could save about □35 crores every year by controlling fruit flies in mango, sapota and cucurbitaceous vegetables. The quality of the products, increase in production and goodwill generated in terms of healthy production ultimately benefited the farmers adopting this technology. Area wide control strategy is not known in our country. Therefore, farmers could also ascertain the actual benefits of adopting such technology which will further motivate them in future.

Fig 4. Fruit fly catches from mango orchard

The effectiveness of low cost IPM technology in controlling infestation of fruit fly disseminated through the RKVY project has brought tremendous behavioral changes in the attitude of fruits and vegetable farmers. Now farmers have adopted this technology which has resulted in export oriented organic production of fruits and vegetables.

In terms of cost-benefit analysis, an estimated benefit of □81,840/ha is achieved by spending merely □ 350/ha. Thus, implementation of fruit fly technology in over 6000 ha of mango could have benefited the farmers to the tune of about □49 crore, while in case of cucurbitaceous vegetables, the fruit fly damage was minimized up to 2.5 to 4.6 per cent by using the technology while the damage was 19 to 32 (30.50) per cent in untreated fields. Thus, more than 85.0 per cent damage due to fruit fly was reduced resulting in yield increase of 27 per cent. Ultimately an estimated benefit of □ 26,250/ha in bitter gourd and □ 39,350/ha in bottle gourd was attained by spending only □ 550/ha. Thus, implementation of fruit fly technology in over 447 ha of cucurbitaceous vegetables crops benefited the farmers to the tune of □1 crore.

Presently, a plan project on "Area Wide Management of Fruit Fly through community approach using Nauroji Stonehouse Fruit Fly Trap in South Gujarat" is running for the cluster-based adoption of this trap for quality food production among farming community.

Fig 5. Inventor of the technology: Dr. Z. P. Patel, Vice Chancellor, NAU, Navsari

Table 1. Selling data of Nauroji Stonehouse Fruit Fly Trap

Year	Nauroji Stonehouse Fruit fly Trap	
	Total sold out (No.)	Total Revenue (INR)
2019-20	12000	6,60,000/-
2020-21	10786	5,93,230/-
2021-22	10200	5,61,000/-
2022-23	13428	8,05,680/-
2023-24	14490	8,69,400/-

Conclusion

In the increasing acceptance of organic agriculture where chemicals are not permitted has led to a quest for the scientific community around the globe to either formulate or suggest safer alternatives for controlling fruit flies. Since the fruit flies are very mobile insects with good flight capacity their spread is also very high and their control is difficult. Hence, area wide adoption of management strategy can only be useful against fruit flies through fruit fly trap which ultimately reduce the pest population for better quality fruits and vegetables.

Implications and Strategic Recommendations

The Nauroji Stonehouse Fruit fly trap is recommended four traps per acre or 10 traps per hectare in mango, sapota, guava and cucurbit vegetable crops.

References*

1. Bansode G.M. and Patel Z.P. 2018. Study on level of awareness about fruit fly menace in farmers of south Gujarat. International Journal of Chemical Studies 6(5):1206-08
2. Bansode, G. M. and Patel, Z. P. 2020. Effect of weather parameters on population fluctuation of guava fruit flies, *Bactrocera Spp. International Journal of Chemical Studies,* **8**(5): 1445-1448.
3. Patel Z.P., J.M. Stonehouse, A. Verghese and J.D. Mumford. 2005. Roles of lures and insecticide in male fruit fly annihilation. Pest Management in Horticultural Ecosystem,11(2):131-132.
4. Patel, S. R., Patel, Z. P., Patel, K. M. and Shinde, C. U. 2021. Surveillance of tephritid fruit flies in the sapota orchards of South Gujarat. Insect Environment. 24(1): 124-128 (ISSN: 0975-1963)
5. Patel, Z.P., R.C. Jhala, V.S. Jagadale, D.B. Sisodiya, J.M. Stonehouse, A. Verghese and J.D. Mumford. 2005. Effectiveness of woods for soaked-block annihilation of male fruit flies in Gujarat. Pest Management in Horticultural Ecosystem, 11(2):117-120.
6. Patel, Z.P., R.C. Jhala, V.S. Jagadale, D.B. Sisodiya, J.M. Stonehouse, A. Verghese and J.D. Mumford. 2005. Methyl eugenol and Ocimum sanctum in fruit fly traps in Gujarat. Pest Management in Horticultural Ecosystem, 11(2):126-128.
7. Venkatachalam, V., Garg, S. C., Purkayastha, A., Chopra, S., Reena Saha, Subbarayan, M., Singh, J. and Selina Sen. 2012. Incentivising Agriculture RKVY Initiatives. Government of India Ministry of Agriculture Department of Agriculture & Cooperation PP. 27-32.

* The content of this chapter was prepared in consultation with above references.

9

Eruvaka Kendras - 25 years of Bringing Modern Agricultural Technologies to Farmers in Andhra Pradesh

G. Sivanarayana[1], V. Visalakshmi[2] and B. Mukunda Rao[3]

[1]Director of Extension, [2]Principal Scientist (Entomology), O/o Director of Extension, [3]Principal Scientist (Extension), O/o Director of Extension, Acharya N.G. Ranga Agricultural University, Lam, Guntur, Andhra Pradesh

Email: directorextension@angrau.ac.in

Abstract

District Agricultural Advisory and Transfer of Technology Centers (DAATTCs) popularly called as Eruvaka kendras in telugu were played key role in transfer of ANGRAU technologies viz, Improved varieties of about 500 in all crops, resource saving DDS system in rice, zero tillage maize, sowing with seed drill, agricultural drones, IPM, INM, IDM and ICM practices in all agricultural crops, pest/disease specific management practices, weed management, resource management, dryland agriculture, soil reclamation, value addition to millets, storage methods, etc., in the way of conducting minikits, technology assessment and refinement, large scale demonstrations, capacity building trainings to farmer, farmwomen, rural youth, NGOs, input dealers, line departments staff, field days, method and result demonstrations. Continuously monitor the field as well as weather situations and provides timely guidance and solutions to the farmers in the field of agriculture. The DAATTCs role was more significant during disaster management in erratic weather conditions, by giving on field suggestions as well as to recommend contingency through print, electron and social media. DAATTCs identify the farming situations, farming systems and cropping patterns village level and recommend suitable cropping systems based on mandal plans to make farming profitable. Also maintains district data base including Rythu Seva Kendra wise data to expand the technologies spread. Also identifies the research gaps for providing farmers needs to the research wing, to unable to reorient the research activities. From last four years in-convergence with RSKs the activities reached to grass root level. Due to DAATTCs about 20 per cent (4,00,000 ha) of paddy area was diverted towards direct sowing, rice, groundnut, blackgram, greengram, redgram and sugarcane varieties released by ANGRAU account for 87, 95, 50, 48, 72, 62 per cent of the total area respectively in Andhra Pradesh. ANGRAU varieties contributing 145.89 lakh tonnes produce with Rs. 26,740 crores from total produce per year.

Keywords: *District Agricultural Advisory and Transfer of Technology Centers (DAATTCs), Rural Awareness Work Experience Programme (RAWEP), Rythu Seva Kendras (RSKs) convergence.*

Introduction

The mandate of the University under section 30 of the Act of 1963 is for establishment of Extension services for the purposes of under taking extension programs, dissemination of information for solving of problems relating to agriculture and domestic fields, developing the interests of youth in agriculture etc., The first extension service of APAU popularly known as District Extension Education Programmes were introduced in 1967 in Hyderabad district and were subsequently extended to Chittoor(1968)and Guntur (1970) districts. National Demonstrations Scheme (NDS) was established during 1970-71 in the University in four locations. In commemoration of the Golden Jubilee Year of ICAR, a Lab to Land Programme was launched by the APAU in June,1979. In order to improve the productivity of pulses and oilseeds in different regions of the state, two extension education centres were sanctioned by the ICAR at Ananthapur and Rastakuntubai which started functioning from December,1980 and December,1981 respectively.

Reorganization of the University Extension work followed the reorganization of research activities of the APAU and implementation of Training and Visit System by the Department of Agriculture to streamline the setup of the field extension program of the University during 1982. To optimize the utilization of available extension resources, the resourcess of District Extension Education programs, extension services around research stations and NATP were integrated and reorganized to form six Extension Education Units. The extension activities of these units were entrusted with the respective Associate Directors of RARS for integrating research and action activities. The Associate directors of Research in turn were made responsible to the Director of Extension. The Extension Education Units conducted different extension activities like organization of field trials for testing new technologies for their feasibility including minikit trials of pre-released varieties to get feedback to the scientists, introduction of new cropping patterns, demonstration of latest technologies through early demonstrations including educational activities like organization of Kisan Melas training programs for the developmental personnel and the farmers, supervision of Rural Awareness Work Experience program for the students of final year B.Sc (Ag) and H.Sc are also undertaken by them.

After knowing the existing EEUs and T&V system is inadequate to strengthen the staff of Agriculture, Horticulture, Animal Husbandry and Fisheries Departments to keep them updated on modern agricultural practices. In 1998 Acharya N.G. Ranga Agricultural University, formerly known as Andhra Pradesh Agricultural University, started District Agricultural Advisory and Transfer of Technology Centres (DAATTC) in Telegu called as Eruvaka Centers for the first time in the Country, one per district located in the District Head Quarter. In order to give technological back up and end support to the extension agencies of the line departments, the present set up of extension services in the University is reorganised as "District Agricultural Advisory and Transfer of Technology Centres (DAATTCs) consisting of multi-disciplinary scientists.

About the Technology Addressed

In 1998 Acharya N.G. Ranga Agricultural University, formerly known as Andhra Pradesh Agricultural University, started District Agricultural Advisory and Transfer of Technology Centers (DAATTC) in Telugu called as Eruvaka Centres for the first time in the Country, one per district located in the District Head Quarter. In order to give technological back up and end support to the extension agencies of the line departments, the present set up of extension services in the University is reorganised as "District Agricultural Advisory and Transfer of Technology Centres (DAATTCs) consisting of multi-disciplinary scientists with the following objectives and mandate.

1. To Develop Database of the District: - Farming situation wise, mandal wise cropping plans were prepared for the district which helps the farmers and extension workers in developing cropping plans based on soil and water requirement. Preparation of Contingency crop plan.
2. Technology assessment and refinement: To Assess and refine technologies generated by the Research stations for suitability to different farming situations in the district.
3. To conduct Diagnostic Surveys periodically: To identify field problems and provide suitable solutions in the field itself to the farmers, where scientists of ANGRAU and Officers from Department of Agriculture and Horticulture are involved.
4. To organize and participate in Kisan Melas, Rythu Chaitanya Yatras, Polam pilustondi, Chandranna Rythu Kshetralu, Rythu Sadassus and organizing exhibitions

5. To organize training programs to Extension functionaries and practicing farmers and participate in training programs organized by the line departments as resource person
6. To maintain an information center at DAATT centre with need based information materials and audio visual aids
7. To render Agro- advisory services to the farmer visiting to the centre.
8. To develop linkages with all the line departments NGOs working for promotion of Agriculture in the district.
9. To Coordinate with AIR, TV and print media for dissemination of information.
10. Development of literature and extend technical support to department of agriculture and other line departments

Extension Management Practices

1. Initiatives

- Varieties of high yielding crops that resistant to major pests and diseases, varieties of crops that support the farmer in any soil and weather conditions
- Photo in sensitive varieties of crops that yields throughout the year, different cropping patterns and cropping systems that can give profit even in adverse weather conditions, introduction of exotic crops and varieties(dragon fruit, cluster bean).
- Integrated crop/pest/disease management tactics, methods of cultivation/ practices to improve the health of soil, water and climate, machinery and tools to reduce labor and overcome labor shortages and improve agriculture, pricing policies to helps to convert agriculture into a business mode, market information, modern information from time to time through line departments, through media, through trainings, direct access to farmers from these Eruvaka Centers has helped to keep University research continuously in the forefront of farmers.
- Taken part in development of Mandal level crop plans, giving contingency crop plan during the time of season failure, disaster management to guide the farmers in saving the crop. Popularization of biofortified varieties to provide quality food to the people. Played important role in success of Polambadi program, implemented by Department of Agriculture.

Farm Radio – www.farmradio.in

Farm radio is the first of its kind online radio initiated by DAATTC, Anakapalle of ANGRAU. Farm radio gives freedom to pod cast what you want and when you want it, with better sound quality to listeners. The services of farm radio to the farmers include ready availability of audio files (podcast) of ICM in groundnut, pests and diseases of rice, maize, sugarcane, cotton and apiary we made available.

Enhanced Linkages with line Departments through Convergence Activities

Linkages with Raitha Samparka Kendras (RSKs)

The Government of Andhra Pradesh has established 10,778 RSKs to offer service to farmers at their doorstep. The RSKs are playing an important role in providing quality seed and fertilizers and also conducting e-cropping for every crop. The Assistants who are working at RSKs should know the latest technologies in agriculture and allied sectors for working effectively at field level. ANGRAU has set up an effective and dynamic research and extension system in Andhra Pradesh for catering the needs of Extension Functionaries and farmers. In this direction ANGRAU has conducted different programs for VAAs of RSKs.

1. **Capacity Building Programs:** From 2020, training programs (2389) were organized with the participation of 86,265 RSK staff including VAAs and VHAs on productivity enhancement of crops, soil health and fertility management, Integrated Pest and Disease Management, Resource Conservation Technologies including Drone technology, Beekeeping, seed production at farmer level to increase farmer net income, commercial production of vegetables, processing and value addition of millets, livestock production and management and protective cultivation. These trainings were imparted to upgrade knowledge and skills of staff of RSKs.
2. **Diagnostic Field Visits:** The changed climatic conditions, cropping pattern and varietal distribution influence the biotic and abiotic stresses on crops. Hence there is a need to monitor the fields continuously for timely identification of the problem to initiate mitigation measures to reduce the crop loss and input wastage. A total of 3475 diagnostic visits were conducted benefiting 73806 staff of RSKs and farmers from last 3 years.
3. **Polambadi:** Farmers Field School (FFS) is a group based adult learning approach that teaches farmers how to experiment and solve problems independently. During last three years, ANGRAU has taken massive steps in popularizing FFS among 32,926 VAAs and farmers by covering 2828 RSKs. Polambadi is like a hands on training to the farmer for

adopting best management practices in the way of correct identification of insect/disease/weed/deficiency symptoms to take appropriate measure, to understand the role of bio-agents, field condition, crop and pest sensitive stages.

4. **Interaction Programs:** A total of 2038 Scientist-Farmer-RSK interaction meetings were conducted benefiting 47053 RSK staff.
5. **Other Extension Activities:** Social media tools such as Whatsapp offer new form of disseminating farm extension information. ANGRAU has created 1330 Whatsapp groups. 54254 VAAs have been covered through these Whatsapp groups and other digital platform. Mandal wise agro-met advisory bulletins (twice in a week), are being provided to VAAs through mobile apps and messages.
6. **Field Days:** A total of 216 field days were organized at 960 RSKs to show the results of field implemented technologies.
7. **Scientist's Participation for Content Recording in RSK Channel:** Padipantalu Studio and Padipantalu Channel at Integrated Call Center, Gannavaram were maintained by Department of Agriculture with the support of Agriculture, Horticulture and Veterinary Universities to provide timely solutions to field level problems. A total of 264 No. of Programmes recorded and broadcasted with Scientists of ANGRAU.
8. **Linking of RAWEP Students with RSKs:** Rural Awareness Work Experience Programme is a part of Academic activity to B.Sc. Ag. course students, during this period 2009 no. of students surveyed 4022 RSKs activities to learn the ground level extension and farmer welfare programs though Department of Agriculture.

Annapurna Krishi Prasar Seva (AKPS)

An Interactive Information Dissemination System (IIDS) is an integrated model to address the problems of farmers by using ICT applications. IIDS is a web, mobile and IVRS based applications, where the farmers are required to be registered with their farm and other details. The expert would provide the personalized solutions based on the queries of the farmers and his farm profile. There is a mobile (Toll free No. 1800 425 3141) interface at front end and web interface at the back end. Data will be transmitted through voice, text, images and videos from both ends (farmers to expert and back) i.e. it would allow farmers to send images / videos of the field along with their queries by using a smart phone.

Benefits to the Farmers through IIDS

- Farmer can get personalized advice on agriculture, horticulture, Animal Husbandry and fisheries from their respective DAATTC on Toll Free number (1800 425 3141).
- Farmer can record their queries 24x7 through Toll Free Number.
- They can get SMS alerts on raised query and solved queries from their respective DAATTC.
- Farmers can receive text and voice messages in Telugu on their mobile from the respective DAATTC.

IIDS-AKPS is one of the most useful tools for dissemination of agricultural information to the registered farmers at user desired mode and time and thereby play a greater role in enhancing efficiency of extension services. The results of the study showed that the majority of the respondents preferred text messages over voice SMS. The growth of mobile market and subscriber penetration is an opportunity to the scientists, development departments, NGOs and IT specialists to diffuse the information by ICT initiatives and mobile technology at a cheaper cost. (Lakshmana *et. al.* 2019)

Reach Every Panchayat

Reach every Panchayat is a unique program formulated and implemented by ANGRAU with a goal of reaching every Panchayat of the state to disseminate improved technologies developed by the University. One key informant farmer is identified in each Panchayat who will influence other farmers' decisions in farming. The Key Informant Farmer along with Sarpanch of the Panchayat are trained and oriented with the best management practices, critical interventions for increasing the productivity of major crops grown in that area, awareness on government schemes and ICT applications. They were provided with prestigious publication of the University Farmer's Almanac Vyavasaya Panchangam and subscription to vyavasayam telugu agriculture magazine published by University for one to place them in the village panchayat library to facilitate their access to the farmers in the village.

Student READY (Rural Entrepreneurship Awareness Development Yojana)

The DAATTCs have been involved to guide the B.Sc. (Hons.) (Ag.) final year students of Agricultural Colleges to provide practical training and experience for one semester by keeping them in villages and through the attachment of one host farmer through RSKs to each student.

Innovative Farmers Network (IFN)

The innovative Farmers Network was initiated by ANGRAU in 2012. One innovative farmer among the five farmers felicitated by the DAATTC during the Foundation Day Celebrations was identified as Coordinator of the Innovative Farmers Network. The Coordinators of network were provided with technology support by the DAATTCs to update their knowledge and skills, who in turn need to share their skills and knowledge to other farmers of the network (30 members) in the district. The main objective of this network is to promote farmer to farmer extension.

Partnership Activities of ANGRAU

- Reliance Foundation: Since its inception in 2013, the Reliance Foundation Information Services (RFIS) program closely working with ANGRAU with aims to provide critical information and linkages using various communication mediums to poor households. Information is disseminated through audio and dial out conferences, local cable TV Live TV Phone in programs, TV scrolls on daily basis. Agro-weather News Bulletins, voice advisories and text SMS, Jio Chat, Whats App and field-based programs and trainings. This enables the farmers to make better decisions in the areas of livelihood, health and disaster preparedness. The broadcast themes include agriculture, horticulture, fisheries, health, employment, micro –enterprises and skill building. The toll free number of Reliance Foundation was 1800 419 8800.
- SERP under APRIGP: In order to upgrade the knowledge and skill of members of the FPOs assisted by the Society of Elimination of Rural Povery (SERP), the University had entered MOU with SERP and implementing the project entitled "Collaborative strategies of ANGRAU-SERP in enhancing the livelihood of small and marginal farmers, nutri and Hygiene entrepreneurship promotion in Andhra Pradesh" under Andhra Pradesh Rural Inclusive Growth Project (APRIGP).

Other Extension Activities

The farmers were imparted knowledge on various practices through TV programs and radio programs. For faster dissemination of information watts up groups were created, though this text messages and voice messages were sent to farmers and extension workers of line departments. Field days, method demonstrations, result demonstrations farmers were conducted as a part of mass communication efforts. The Acharya N.G. Ranga Agricultural University has been organizing Kisan melas at various research stations and colleges throughout the state to create awareness and to educate the farmers about latest farming technologies and development. Kisan Melas provide an opportunity to gain first-hand information on the latest technologies, live demonstrations, informative agricultural exhibitions, interaction with the scientists, input agencies and inculcate the habit of visiting research stations frequently for exposure and timely advices.

Reach

The DAATTCs are meeting the farmers needs and in convergence with department of agriculture, line departments and NGOs time to time. One of the major activites of DAATTCs is to assess and refine technologies starts with testing the advanced lines of crop varieties in minikits to assess its performance in different field conditions in all agro-climatic zones. Till now 2,519 minikits were organised, and the feedback information was useful to release 494 crop varieties from ANGRAU. The research findings were tested through OFTs, from these the successful ones were popularised through FLDs, field days, capacity building programs. Till date 1423OFTs were organised 912 successful OFTs demonstrated through FLDs. 7249 training programs were organised to extension personnel, farmers, farm women, rural youth. Also popularised through mass media through 1020 All India Radio programs, 2571 Television programs, send 8605 messages through IIDS, whatsApp etc., Conducted 30,857 diagnostic field visits alone, in a team with line departments to identify biotic, abiotic stresses and to give suitable suggestions to mitigate that problem timely to recover higher yields with limited cost of cultivation and making agriculture profitable. Also act as liaison centre for availability of quality inputs (Major seed) and improved technologies to farmers directly as well as extension persons. Their role is most prominent in erratic weather conditions like cyclones, drought, untimely rains to guide farmers to minimise damage and to switch on to alternate crops or package to thrive well under unfavourable situations.

Rural Awareness Work Experience Programme (RAWEP) for final year students of Agriculture B.Sc.(Ag.) is an excellent program to learn by doing the things, implemented by these centres, till date 6,775 students were guided by DAATTCs.

Output

As a result of organisation of OFTs followed by FLDs and capacity building programs by DAATTC, Srikakulam on zero tillage Maize spread in 15500ha, Direct sown paddy in 1,00,00ha, new varieties of paddy and pulses in 1,00,000ha which results reduction of cost of cultivation along with increased productivity and net returns. As a result of organisation of OFTs followed by

FLDs by DAATTC, Vizianagaram since 2017 on zero tillage tractor drawn marker and post emergence herbicide and recommendation through interaction and Mass media on zero tillage Maize the technology spread in 3600 ha with economic contribution of 234 lakhs, Drum seeder paddy the technology spread in 2358 ha with economic contribution of 198.072 lakh. sowing with seed drill the area spread in 1500 ha with economic contribution of 45.0 lakh.

Fall army worm management in Maize farmers able save the crop in 7000 ha. Variety MTU-1121 spread in 73,000 ha area (60% district area) as on 2019-20 in Vizianagaram district. Due to the technology transfer activities of DAATTC, Vizianagaram Maize growing farmers realized 40 per cent increase in net income due to increase in grain yield by 7.72 per cent with reduction in cost of cultivation by 11.12 per cent (Srinivas Rao *et.al*). As a result of organisation of OFTs followed by FLDs by DAATTC, Peddapuram Newly released variety MTU 1318 is gaining importance among farmers & occupied >10,000 ha in the district and slowly replacing the popular variety SWARNA. MTU 1140 suitable for stagnant flooding is popular in Konaseema area. Timely intervention by DAATTC Eluru during rabi season, loss due to BLB, Sulphide injury and stem borer was minimized in rice and Bacterial stalk rot in Maize. DAATTC, Ongole by its activities popularised YMV resistant TBG104, GBG-1 in 22000ha, *Maruca* management in 65000ha, foliar sprays in Bengal gram in 58000ha to get higher yields and for drought mitigation, yield increase by 15-20% in major crops by adopting ANGRAU varieties and introduction of foxtail millet as proceeding crop to Bengal gram in 25000ha in Prakasam district. DAATTC, Nandyal popularised mechanical harvesting in Bengalgram and dry land technics. DAATTC, Puttaparthi popularised effective management module for PSND in Groundnut. In the Vizianagaram district the ICT & mobile applications are enabled to reach more no. of farmers who geographically dispersed over the district by simple text message from AKPS is able to reach 6530 registered farmers at a time and what app message is reaching 250 respondents with in few seconds. ICT enable the farmers to share the photographs of pest/disease infested plant with extension scientist and readily receive a solution for their problem by sitting at farm (Lakshmana *et al.*).

Outcome / Impact

Andhra Pradesh grows paddy on more than 21 lakh hectares with transplanting being the predominant cultivation practice. Encouraging paddy farmers to adopt more water efficient practices such as Direct Seeded Rice (DSR) and Alternate Wetting & Drying (AWD) reduces methane emissions, water usage, and improves soil health. It reduces the cost of cultivation and improves farmer profitability and also generates carbon credits. With the efforts of DAATTCs about 20 per cent(4,00,000 ha) of paddy area was diverted towards direct sowing, rice varieties released by ANGRAU account for 87 per cent of the total area, groundnut varieties accounted for 95 per cent of the total area covering 7.80 lakh ha., Blackgram varieties released by ANGRAU grown on 1.89 lakh ha, accounting for 50 per cent of the total blackgram area, Greengram varieties grown on 0.413 lakh ha, accounting for 48 per cent of the total greengram area, more than 72 per cent of the state's redgram cultivation area is occupied by ANGRAU varieties, accounting for 1.84 lakh ha, ANGRAU varieties occupy 62 per cent of sugarcane area grown on 50.01 thousand ha in the State of Andhra Pradesh. ANGRAU varieties contributing 107.51lakh tonnes produce worth 20905 crores from cereals, 2.786 lakh tonnes produce with 1808 crores of Rs. from pulses, 5.25 lakh tonnes produce with 3100 crores of Rs. from oil seeds and total 145.89 lakh tonnes produce with 26740 crores of Rs. from total produce for the year 2022-23. (CARP, ANGRAU, 2022-23).

Conclusion

The District Agricultural Advisory and Transfer of Technology Centers completed silver jublee year of establishment, and played crucial role for faster dissemination of latest agricultural technologies from ANGRAU in convergence with Rythu Seva Kendras, line departments, NGOs, FPOs. They are the face of the University for farmers as well as line departments. With their continuous efforts not only popularised latest high yielding pest/disease/ disorder resistant varieties but also the significant technologies like Dry direct sowing, wet direct sowing, aerobic rice, zero tillage rice, rice fallow sun hemp seed production, yellow mosaic management in pulses, BPH management in rice, PBW in cotton FAW in maize, popularization of biofortified and improved varieties in millets, *B.t* cotton, etc.

Implications/ Strategic Recommendations

It is an excellent system to reach out to farmers and to strengthen the line departments, NGOs, FPOs, Water user association staff regarding modern agricultural technologies and to provide contingency crop plans with in short time under fragile agricultural system to stabilize farmer's income. Hence this system can be replicated in other parts of the country to take its advantage in the field of farming.

References

1. Centre for Agriculture and Rural Development Policy Research- 2022-2023. Regional Agricultural Research Station, ANGRAU, Lam, Guntur.
2. Lakshmana Kella, Sandhya Rani Y., Harika G. and Punna rao P. 2019. Use of Information and Communication Technologies for Diffusion of Information to The Farming Community-A Successful Case in Vizianagaram District, India. *International Journal of Agriculture Sciences*, 11(18): 9022-9024.
3. Lakshmana Kella, M. Swathi, K. T. Rao and P. Rambabu. 2022. Use of Information and Communication Technology for tranfer of technology- The Impact on adoption of Improved agricultural technology. *International Journal of Plant Sciences*, 17(2):241-243.
4. Srinivasa Rao M. M. V., K. Lakshman and Roy G.S. 2016. Zero tillage method of maize cultivation is a boon to the farmers in Vizianagaram district of North Coastal Zone of Andhra Pradesh. *Progressive Research-An International Journal*, 11(Special IX):6504-6507.

10

Innovative Methodologies in Agriculture Extension Adopted by KVKs Under UAS, Dharwad

Mouneshwari R. Kammar[1], S.N. Jadhav[2], P. Nagaraju[3], R. Basavarajappa[4], S.A. Gaddanakeri[5], and R.B. Belli[7]

[1,2,3,4]Directorate of Extension, [5,6]Associate Director of Extension, University of Agricultural Sciences, Dharwad

Email: de@uasd.in

Abstract

This paper presents a compilation of innovative methodologies adopted by the Directorate of Extension through Krishi Vigyan Kendras (KVKs), Agricultural Extension Education Centres and Staff training units under the University of Agricultural Sciences, Dharwad over the past five years. This paper throws light on It highlights the use of both conventional technologies transfer methods and ICT techniques to enhance farm productivity and effectively reach the farming community. A systematic approach is employed by KVKs to identify and prioritize agricultural challenges at the district level, developing tailored solutions that leverage both local and national resources.

Keywords: *Agriculture, Innovative Methodologies, ICT tools, Krishi Vigyan Kendras, Transfer of Technologies*

Introduction

University of Agricultural Sciences, Dharwad is one of the premier agriculture Universities in India has made its remarkable journey in the field of Teaching/ Research / Extension since its inception, UAS, Dharwad has got six Krishi Vigyan Kendras sponsored by Indian Council of Agricultural Research. Which have adopted various outreach methodologies for the farming community. Apart from conventional methods of technology transfer i.e., technologies in pipeline (farm trails) conducted in farmers fields for their acceptance, front line demonstration, on farm testing, PRA, diagnostic field visits etc. KVKs have employed various techniques making use of ICT which have proved to be effective in enhancing the productivity of farms (Medhi *et al.,* 2017), reaching farmers quickly and effectively. The present paper is a compilation of innovative methodologies adopted by KVKs under UAS Dharwad during last five years.

As it is mandate for all the KVKs to prepare action plan before the onset of every year during March-April, vigorous exercises are done by KVKs to prioritize the problems faced by farmers in the particular district. These problems are planned to tackle the location specific technologies developed in that zone. If the technologies under FLD banner are not available in particular technologies developed elsewhere in NARS are borrowed and tested for both its sustainability under that agro-ecological system. During this process of prioritizing the problems, implementation and monitoring, KVK adopt various methods given as discussed below.

Krishi Mela

Agriculture College Dharwad has initiated field-day program during 1956-57 to solve the problems / constraints of the farmers by involving faculty scientists, students and staff of the college for few years. Later with financial assistance from State and Central Government developmental departments, these field days were converted to Krishi Mela due to increase in participation of farmer number from thousands to lakhs. Since last three decades UAS, Dharwad through it's good quality education, high quality research, seed production and extension activities got recognition in national and international level. Under extension activity UAS, Dharwad regularly conducting Mega Krishi Mela (4 days) during 2nd fortnight of September every year More than 20.00 lakhs farmer's / farm women/ students/ extension functionaries are participating. The number of farmer's foot fall in the krishimela conducted is as detailed below:

Table 1. Footfall in Krishimela

S. No.	Year	No. of Participants
1	1987	5,000
2	1988	5,000
3	1990	7,000
4	1991	14,000
5	1992	14,000
6	1993	10,000
7	1994	13,000
8	1995	15,500
9	1996	21,830

10	1997	22,600
11	1998	29,200
12	1999	35,000
13	2000	40,600
14	2001	45,000
15	2002	45,000
16	2003	25,000
17	2004	1,13,000
18	2005	1,50,000
19	2006	4,68,000
20	2007	51,000
21	2008	5,85,000
22	2009	5,00,000
23	2010	7,40,000
24	2011	10,91,000
25	2012	10,09,000
26	2013	10,50,000
27	2014	11,00,000
28	2015	11,05,000
29	2016	16,00,000
30	2017	16,05,000
31	2018	10,00,000
32	2019	5,10,000
33	2022	15,37,000
34	2023	10,35000
35	2024	21,28,000

* Krishi Melas were not conducted during 2020 and 2021 because of flood and covid-19 respectively

Interaction during Krishi Mela

Diagnostic visit either for Identification by KVKs or in Collaboration with Departments

The field problems are reported by the agriculture and line departments, and horticulture departments, and a committee is formed at the university level including scientists of different disciplines. A visit is made and recommendations are passed to the department, thus based on the recommendations, the chemicals and necessary inputs are indented by the department. Eg., Diagnostic field visit to root grub in sugarcane, nutrient deficiency in turmeric Chrysanthemum blight management, citrus canker, etc.

District Agromet Advisories (DAMU)

UAS Dharwad has got District Agromet Units sponsored by ICAR through IMD Dept. These units get every minute data on weather forecasting by IMD. This data is converted to advisories and taluka-wise data of weather forecasting is disseminated through weekly bulletins which advisese farmers on weather forecasting for next four days and the corresponding agricultural operations to be undertaken by the farmers. However, the data is being received by the stations through well-established weather forecasting stations.

One District One Product–initiative

Upon the declaration of the theme one district-one product by the central government each district was tagged with one product. KVKs have started several initiatives in popularizing this concept. The product popularization through Festival was one among them. Eg. Bagalkot district has jaggery as its product under ODOP. So an event in name Jaggery Fest was organized bringing in all the stakeholders in the line of production of marketing and value addition. This event was organized involving district administration, Member of Parliament and Legislative assemblies, department of agriculture and others. Eminent speakers in the field of value addition and marketing interacted with farmers and thus popularized the product.

Jaggery Fest

Through ICT Media

Information Communication Technology (ICT) has embedded in the lives of the farmers. Starting with phone call, WhatsApp group, youtube channels, Instagram messages, twitter handler, are being extensively used by the Krishi Vigyan Kendras. KVKs created whatsapp groups of CIG/FIG or the trainees of different programs to discuss the field problems and provide the solutions. In some instances, KVK staff are added as members in WhatsApp groups created by other stakeholders which are acting as an important means to reach large number of farmers. Krishi Sakhis/ SHGS who report the problems on groups along with photographed query. Such queries are answered by the experts. In addition, creating YouTube channels, documenting success stories are also done by KVK. QR codes are created for each technology, so that farmer can easily scan through and get information of the technology. KVK Hanaumanmatti is an example of creating these QR codes for technologies. On an average each KVK has minimum 15-20 whatsapp groups with 150-250 members

SWOT Analysis

Though not an innovative concept, SWOT analysis is used extensively in the preparation of action plans of KVKs. It is a strategic planning tool helps in identifying strengths, weaknesses, opportunities and threats. KVKs make use of this tool to assess the resources to improve the weaknesses and utilize the opportunities and alleviate threats so that they can better make use of existing resources to inform the farmers. Available staff strength, cropping pattern, water resources availability of marketing, opportunities for value addition are few factors considered in SWOT analysis, while preparing action plan.

Capacity Building of Farm facilitators of Bhuchetana Scheme

International Crop Research Institute for Semi-Arid Tropics (ICRISAT) had sanctioned Bhuchetana scheme to Karnataks State Department of Agriculture with a provision for human resource in the name of Farm facilitators. Farm facilitators are working at village-level as extension workers. Training these people on agriculture and allied sectors was helpful in transfer of technologies. Solving the farmer's field problems through Farm facilitators is definitely time saving and cost saving. Problems can be solved immediately. Programs can be easily organized with help of farm facilitators through arrangements and mobilizing farmers.

Kisan Mobile Advisory Services (KMAS)

Each KVK has got the list of contact farmers enrolled for receiving the Kisan Mobile advisories. The registered farmers will be sent with an advisory tip on various crop-specific technologies every week. Thus, each KVK has around 10000-50000 registered farmers with them. Initially weekly messages were sent to them based on the season and major crop of the district. However, after the restriction on the number of messages by the govt. these farmers are receiving messages through whatsapp groups.

Everyday Agri Advisory through Messages

During COVID-19, the KVKs were entrusted with the work of reaching to farmers and facilitate marketing of their produce. Thus, each KVK along with the team of scientists visited majority of the villages and listened to their problems, facilitated a platform by giving the details of buyers and sellers - to market their produce. Daily advises were sent continuously even upto three years even after COVID, as this platform was appreciated by most of the farmers.

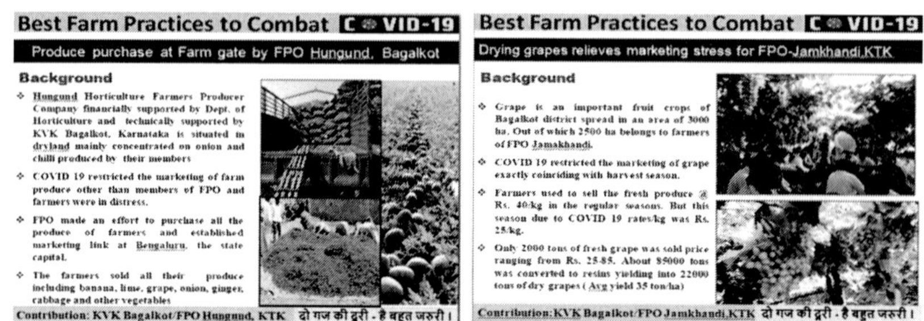

SMS service for farming community

Under Kisan Mobile Advisory scheme (mKISAN), daily SMS alerts were issued on various agricultural technologies namely weather forecast, disease forecast, market information, pest & disease management, etc. The service is also being used as a medium to send information on important trainings and other programs to the members of the Farmers Clubs and SHG network under the KVK. These messages were sent twice a week initially, later these programs were merged with private partners namely reliance radio tips, community radio stations operating in the district and All India Radio on a regular basis.

Creation of Technical Agents and Leadership Development for Transfer of Technology

The KVK linked with NGO, SHGs, ATMA and development departments to transfer the technology. The ground level workers are identified by the KVKs and are trained in various aspects of crop management. These technical agents reached to farmers in solving the problems. ATMA staff under department of agriculture, are the main workforce in serving the farming community. They are regularly trained by the KVKs and SAMETI also.

Pest Surveillances

The pest and disease outbreak reported by the agri and allied departments, or outbreak during heavy rainfall/drought, or during field visits by the scientists of KVK, will be surveyed pest surveillance. Upon the formation of a committee by the university suitable recommendations will be issued for that particular system through department, messages, popular articles, print and electronic media.

Rapid Roving Survey

Number of scientist teams are formed by the university to conduct pest surveillance at various points simultaneously in order to get the comprehensive magnitude of the problem in case of severe outbreak of pest and diseases. Eg. Helicoverpa *armigera* outbreak in Pigeon pea and chickpea, Fall army worm outbreak in maize and sorghum.

Direct Phone in Programs and Tips to the Farmers through All India Radio

Based on the season and crops messages are being sent to AIR Vijayapura which is being broadcasted by AIR regularly which includes crop production technology, Soil and water management, Pest and disease management, etc.

Broadcasting of Success Stories through AIR and Television

Success stories of progressive and awardee farmers were broadcasted through AIR and Television to motivate other farmers to take up agriculture as a profitable enterprise. In this program progressive farmers share their experiences and gave suggestions to others farmers. These farmers were also recognized by the department and are incentivized to mobilize more number of farmers in their surroundings.

Involving Progressive Farmers as Master Trainers in KVK Training Programs

On and Off campus training programs are being regularly conducted in KVK wherein progressive farmers are being invited as master trainers to share their knowledge and success with other farmers. Here other farmers get an opportunity for direct interaction with other farmers. UAS Dharwad initiated a special programs in this direction in name of Farmers to farmers, capapcity building of army personnel, etc.

Demonstration of IFS Model by KVK through establishing different units

A ready model adopting the principle of recycling, reuse and reduced use of inputs is established at KVKs under UAS, Dharwad These models are based on the location specific and season specific technologies. This models serves as a live specimen for the training and also for the visiting farmers. All the KVKs essentially incorporated vermicompost, nutrigarden, fish cum farm pond, goat and sheep unit, nursery under these models. These are also serving as resource models for livelihood sustenance for the sakhi trainings. This model is used for training or teaching the farmers.

Farmers Field School

Under the supervision of an expert trained in conducting Farmers Field school, every KVK is conducting a FFS based on the problem faced in major crop of the district. During COVID-19 virtual farm schools were also conducted by KVKs on different crops. A farmer will be selected for adopting and new technology and other farmers surrounding his farm will be attending the session's. Scientist act as facilitators for this program.

FPO-KVK Linkage

Farmer Producer Organizations (FPOs) are acting as change agents in agriculture. These FPOs are promoted by various institutions such as state government, central government, department of agriculture, horticulture, IIMR, NABARD, NCDC etc. KVKS are given the responsibility of formation, guidance and follow up of FPOs as Cluster Based Business Organizations (CBBOs) as commendable. This is a program sponsored by KSDH to promote linkages of FPOs with KVKs for technical back stopping. All the KVKs under UAS Dharwad are involved in technology transfer.

Table 2. Details of KVKs under UAS, Dharwad working with FPOs

S. No.	KVKs	Names of FPOs	Address
1	Hanuma namatti	Haveri FPO	Haveri, Haveri (Tq&Dist)
		Bhoomika FPO	Ranebennur, Haveri (Dist)
		Kantesh FPO	Kadaramandalagi, Byadgi (Tq), Haveri
2	Indi	Shri Shantheshwara Horticulture Producers Company, Ltd. I	Indi
		The Indi Savayava Hagu Siridhanya Producer Company Ltd.	Indi
3	Dharwad	Uluva Yogi Horticulture Farmers Producer Company Ltd.	Hubballi
		Kayaka Yogi Horticulture Farmers Producer Company Ltd.	Dharwad
		Dummavad Farmers Producer Company Ltd.	Dummavad, Kalghatgi Taluk , Dharwad
4	Sirsi	Pragati mitra North canara farmer producer com. Ltd,	Sirsi
		Shri madhukeshwara Horticulture farmer producers co. Ltd.	Andagi, Sirsi
		Dodnalli Shambulingeshwar farmers producers com. Ltd.	Dodnalli, sirsi
		Banavasi Raita MItra Jeevidhan Producers com. Ltd.	Banavasi Sirsi Taluk
		Bhagyavidhata Farmers Producers Com. Ltd.	Bilagi, Siddapura Tq.
		Aghanashini Spices Farmers Producers com. Ltd.	Nanikatta, Tyagli, Siddapur Tq.
		Shri Rameshwara Savayava Krushi Privara farmer producers com. Ltd.	Itagi Siddapura Tq.
		Dodmane Laxminarayan Farmers producers com. Ltd.	Dodmane Siddapur Tq
		Bhusiri Framers producers com. Ltd.	Nilkunda, Siddapura Tq.
		Mattigatta Belesi Farmers producers com. Ltd.	Mattigatta,Sirsi Tq
		Raita Kalyana Farmers producers com. Ltd.	Pala, Mundagod Tq
		Nisargadhari Farmers producers com. Ltd.	Mirjan, Kumta Tq
		Kali Farmers producers com. Ltd.	Deria, Joida Tq.
		Sarvajnendra Farmers Producer Company Ltd.	Ummachagi, Yellapur Taluk, Uttara Kannada

		SrideviSeva Farmers Producer Company Ltd.	Kalache, YellapurTq, Uttara Kananda district
5	Vijayapur	Shri Basaveshwar Farmer Producer Organization	Basavana Bagewadi taluk of Vijayapura
6	Bagalkote	Shree Renuka Rait Utpadak Sahakari Sangha Limited	Sector no-35, plot no-28, Buddar complex Navanagar Bagalkote
		Banashri Raith Utpadaka Sahakara Sangha	Aladakatti Cross Tq: Badami, Dist: Bagalkote

Rabi Mela

KVK organizes Mini kisan mela to disseminate knowledge on various crops. During crop harvest of *Rabi* eg. wheat, chickpea, safflower a Mela is being organized in the name of field day/*rabi* mela to showcase the advanced technologies adopted by the KVK/ARS to the farmers. UAS Dharwad is hosting many Melas along with mega krishi Mela. In Vijayapur campus, a relatively big event involving all the stake holders also is organized during first week of January every year.

MCX Price Forecast Ticker Board for Displaying Prices

The Mumbai stock exchange (MCX) company has volunteered to display the daily rates of agricultural commodities in KVKs, so that the visiting farmers can get benefitted through these displays. However, district specific crops were not covered under this, it may have modified according to the local needs. Not only the rates, even the technological solutions to the major field problems can also be displayed.

Integration of Methodologies for Enhancing Pulse Production in the district

The on-farm testing which have been taken up by the krishi vigyan kendras, borrowing technologies from other institutes and converted to front line demonstrations upon the success in on farm testing. These front line demonstrations were again promoted through seed production at KVK farms. Hence it was an integration of methodologies such as FLDs OFTs, seed production under seed hub, trainings, production and supply of trichoderma and vrmiwash.

Mass Campaign about Pest and Disease Outbreak

Mass campaign about outbreak in management of major pest and disease outbreak i.e., Fall Army worm, Helocoverpa, Root grab in sugarcane

Farmers Participatory Seed Production on Collaboration with FPO's and Individual Farmers

Due to heavy demand can University variety the seed availability is a constraint. Hence, farmer's participatory seed production in all the KVKs war proved to be effective. The monitory of seed production was done jointly by KSSC and University Eg. KMP-105 short duration paddy seed production, Hemavati paddy variety recommended for water logging conditions production at KVK Sirsi, Greengram & Blackgram in Dharwad.

Working with CIG/ FIG

Trainings are organized for Co-operative Society members, Capacity building of FPO directors and members, Panchayath President and Vice-President, and members of GPS and taluka Panchayat for effective dissemination of technologies.

Pre-seasonal Campaigns

Pre Kharif (May 1st fortnight) campaign and Pre Rabi campaign (September1st fortnight) are conducted regularly before the onset of sowing for particular season to popularize the varieties and other technologies developed by the University.

Bi-monthly Workshops

Bi-monthly workshops are arranged for every two months with the Joint Director of Agriculture of the respective district to get feedback regarding field situation, field problems and provide solution to the agricultural officers of all the taluks of the district with the help of a team of scientists of UAS, Dharwad and also educate about modern agricultural technologies to be carried out during next two months to attain higher productivity on sustainable basis. Bi-monthly workshops organized during the year 2023-24 given below.

Table 3. Bi-monthly workshops organized during 2023-24 (Zone VIII, IX and X)

S. No.	Title of the Meeting	Date	Place
1	Bi-Monthly Technical Workshop along with KSDA officials	22-06-2023	JDA Office, Belagavi
2	Bi-Monthly Technical Workshop along with KSDA officials	23-06-2023	DATC, Dharwad
3	Bi-Monthly Technical Workshop along with KSDA officials	27-06-2023	JDA Office,Haveri
4	Bi-Monthly Technical Workshop along with KSDA officials	28-06-2023	DATC, Kumta
5	Bi-Monthly Technical Workshop along with KSDA officials	11-10-2023	KVK, Sirsi

6	Bi-Monthly Technical Workshop along with KSDA officials	18-10-2023	DATC, Devihosur
7	Bi-Monthly Technical Workshop along with KSDA officials	20-10-2023	DATC, Dharwad
8	Bi-Monthly Technical Workshop along with KSDA officials	08-11-2023	JDA Office, Belagavi
9	Bi-Monthly Technical Workshop along with KSDA officials	18-01-2024	DATC, Dharwad
10	Bi-Monthly Technical Workshop along with KSDA officials	06-02-2024	DATC, Devihosur
11	Bi-Monthly Technical Workshop along with KSDA officials	08-02-2024	DATC, Kumta
12	Bi-Monthly Technical Workshop along with KSDA officials	18-03-2024	DATC, Arbhavi
(Zone III)			
13	Bi-Monthly Technical Workshop along with KSDA officials	21-22 February, 2023	DATC, Vijayapur
14	Bi-Monthly Technical Workshop along with KSDA officials	07-08 June, 2023	KVK, Hulkoti, Gadag
15	Bi-Monthly Technical Workshop along with KSDA officials	11-12 July, 2023	DATC, Vijayapur
16	Bi-Monthly Technical Workshop along with KSDA officials	14-15 July, 2023	DATC, Bagalkot
17	Bi-Monthly Technical Workshop along with KSDA officials	04-05 October, 2023	DATC, Vijayapur
18	Bi-Monthly Technical Workshop along with KSDA officials	22-23 October, 2023	DATC, Bagalkot
19	Bi-Monthly Technical Workshop along with KSDA officials	09-10 November, 2023	DATC, Bagalkot
20	Bi-Monthly Technical Workshop along with KSDA officials	04-05 December, 2023	KVK, Hulkoti, Gadag
21	Bi-Monthly Technical Workshop along with KSDA officials	17-18 January, 2024	DATC, Bagalkot

Weekly Webinar

Through SAMETI online webinars are conducted to the ATMA staff and developmental department staff on various aspects weekly. This has very good role in bridging the technology gap among the grass roots level functionaries.

Table 4. State-level Trainings during 2022-23 (Online)

S. No.	Name of the Training Programme	No. of Participants	Date
1	Awareness program to Belagavi dist (Belagavi division) Gram Pradhans on Natural Farming	75	04-07-2022
2	Awareness program to Kalaburgi dist (Kalaburgi division) Gram Pradhans on Natural Farming	88	04-07-2022
3	Awareness program to Haveri dist (Haveri division) Gram Pradhans on Natural Farming	63	05-07-2022
4	Awareness program to Bidar dist (Bidar division) Gram Pradhans on Natural Farming	61	05-07-2022
5	Awareness program to Uttar Kannad dist (Karwar division) Gram Pradhans on Natural Farming	59	07-07-2022
6	Awareness program to Raichur dist (Raichur division) Gram Pradhans on Natural Farming	81	07-07-2022
7	Awareness program to Vijayapur dist (Vijayapur division) Gram Pradhans on Natural Farming	55	08-07-2022
8	Awareness program to Yadgiri dist (Yadgiri division) Gram Pradhans on Natural Farming	48	08-07-2022
9	Awareness program to Bagalkot dist (Bagalkot division) Gram Pradhans on Natural Farming	54	11-07-2022
10	Awareness program to Ballari dist (Ballari division) Gram Pradhans on Natural Farming	70	11-07-2022
11	Awareness program to Dharwad dist (Dharwad division) Gram Pradhans on Natural Farming	47	12-07-2022
12	Awareness program to Koppal dist (Koppal division) Gram Pradhans on Natural Farming	76	12-07-2022
13	Awareness program to Gadag dist (Gadag division) Gram Pradhans on Natural Farming	48	14-07-2022
14	Awareness program to Vijayanagar dist (Hospete division) Gram Pradhans on Natural Farming	76	14-07-2022
15	Awareness program to Kalaburgi dist (Sedum division) Gram Pradhans on Natural Farming	52	15-07-2022
16	Awareness program to Belagavi dist (Chikkodi division) Gram Pradhans on Natural Farming	74	15-07-2022
17	Awareness program to Bidar dist (Basavakalyan division) Gram Pradhans on Natural Farming	75	18-07-2022
18	Awareness program to Haveri dist (Ranebennur division) Gram Pradhans on Natural Farming	32	18-07-2022
19	Awareness program to Raichur dist (Lingasur division) Gram Pradhans on Natural Farming	57	19-07-2022
20	Awareness program to Uttar Kannada dist (Kumta division) Gram Pradhans on Natural Farming	72	19-07-2022

21	Awareness program to Vijayapur dist (Indi division) Gram Pradhans on Natural Farming	35	21-07-2022
22	Awareness program to Gadag dist (Ron division) Gram Pradhans on Natural Farming	28	21-07-2022
23	Awareness program to Dharwad dist (Hubli division) Gram Pradhans on Natural Farming	63	22-07-2022
24	Awareness program to Bagalkot dist (Jamakandi division) Gram Pradhans on Natural Farming	44	22-07-2022
25	Awareness program to Vijayanagar dist (Harapana-halli division) Gram Pradhans on Natural Farming	74	25-07-2022
26	Awareness and Usage of Nano urea	204	07-10-2022
27	Production Technology of Mahagani	166	11-10-2022
28	Production Technology of Bamboo	141	13-10-2022
29	Mahila Kisan Diwas	206	15-10-2022
30	Effective Implementation of ATMA Scheme **(offline)**	40	15-17 November, 2022
31	National Mission on Natural Farming (NMNF) workshop **(offline)**	176	17 March, 2023
	Total	**2440**	

Table 5. State-level Trainings during 2023-24 (Online)

S. No.	**Name of the Training Programme**	**No. of Participants**	**Dates**
1	Weather Forecasting -2023-24	298	19-05-2023
2	Livestock Management in Summer	304	26-05-2023
3	Calculation of chemical fertilizers in agriculture crops & use of Nano urea	301	02-06-2023
4	Implementation of ATMA activities in 2023-24 **(Offline)**	202	07-06-2023
5	Importance of Bio fertilizers in Agricultural Crops	294	09-06-2023
6	Use of Solar energy in Agriculture	612	14-07-2023
7	Integrated Management of Wilt Disease in Redgram	3000	19-07-2023
8	Management of Fall Armyworm in Agriculture Crops	671	28-07-2023
9	Introduction of weeds and their management in Agriculture crops	652	11-08-2023
10	Modern Agriculture Implements	683	01-09-2023
11	Drought Management: Suitable Agricultural Technologies	665	15-09-2023
12	Management of livestock and Fodder in drought	601	13-10-2023

13	Secondary Agriculture - Fruit Pulp & Vegetable based Processed Products	590	17-11-2023
14	Secondary Agriculture – Bee farming	600	08-12-2023
15	Drone technology in Agriculture	1000	15-12-2023
16	Secondary Agriculture – Mushroom cultivation and value addition	1402	19-01-2024
17	Production Technology of Bamboo	1200	09-02-2024
	Total	**13075**	

Krishi Shodhane-Sadhane

A unique weekly program entitled 'Krishi Shodhane-Sadhane' was sponsored by University of Agricultural Sciences, Dharwad which is being broadcasted on All India Radio, Dharwad since 17.11.2022 to disseminate latest Agricultural technologies to the farmers developed by the University.

Weekly Interviews

Through All India Radio interviews with agricultural scientists are being broad coasted every Monday.

Krishi Community Radio Station (KCRS)

The first Community Radio Station in India under State Agricultural Universities was established at University of Agricultural Sciences, Dharwad campus on May 17, 2007 with Frequency of 90.4 MHz. The main theme is to empower the farming community. KCRS covers 41 villages within 15-20 kms radius. To have more coverage area, KCRS Radio Mobile App was launched during June, 2022. The journey of Krishi Community Radio Station has completed 16 successful years. Every day it broadcasts agricultural programs related to agriculture, horticulture, animal husbandry, health programs, programs on special days for 6 hours from morning 6 am to 9 am, and same is repeated on the same day evening 6 pm to 9 pm in Kannada language.

The above listed methodologies are extensively used by the Directorate of Extension; the list is not a comprehensive. Above methodologies are also supported with individual phone calls to farmers, home visits, field visits, specimens brought to KVK/ University. However, individual farmer has their unique way of getting the information needed in farming.

Reference

1. Medhi, S., Singha, A.K., Singh, R. and Singh, R.J., 2017. Effectiveness of training programs of Krishi Vigyan Kendra (KVK) towards Socio-economic Development of Farmers in Meghalaya. *Economic Affairs*, 62(4): 677-682.

11

Introduction of Apical Rooted Cuttings (ARC) Potato for Seed Tuber Production: A Novel Approach for Profit Maximization in Hassan District

Rajegowda[1], Y.N. Shivalingaiah[2], N. Pallavi[3] and Ashoka Doddamani[4]

[1]*Sr. Scientist & Head, ICAR-KVK, Hassan,* [2]*Director of Extension,* [3]*Scientist (Crop Physiology), ICAR-KVK, Hassan,* [4]*Scientific Officer, Directorate of Extension, UAS, Bengaluru*

Email: deuasBengaluru@gmail.com

Abstract

Potato is an important commercial crop and Hassan district is the place where potato is extensively grown. Even though many districts are producing large quantities of potatoes, the state average yield still lags behind compared to other states. Initially, the area under potato cultivation was 40,000 ha, but now it's been reduced to 10,000 ha, due to various reasons viz., non-availability of quality planting materials, untimely supply, poor germination and disease outbreak. Hence, providing quality tubers in time is a need for the hour. In this regard, Apical Rooted Cuttings (ARC) of potato produced from tissue culture technology has changed the scenario of potato cultivation in Hassan district in terms of both income as well as employment generation. Keeping in view of the above facts, ICAR- Krishi Vigyan Kendra, Kandali, Hassan has conducted trials on evaluation and performance of ARC potato for quality seed tuber production in association with Horticulture Research & Extension Centre (HREC) Somanahalli Kaval, Department of Horticulture, Hassan, German Innovation Center (GIZ/GIC), International Center for Potato (CIP) and Education and Training Consultancies (ETC) Bengaluru. The experiment was carried out by adopting Randomized Complete Block Design (RCBD) with twenty replications and four treatment combinations with different genotypes. Results revealed that ARC treatment combinations viz., Kufri Himalini-ARC recorded significantly higher percent establishment (97.88 %), maximum plant height (90.49 cm), no. of leaves (82.03), no. of tubers per plant (8.58), total tuber yield per plant (252.74 g) and total tuber yield of 19.53 t/ha followed by Kufri Jyoti-ARC. Further, higher B : C ratio was obtained from Kufri Himalini-ARC (3.5) followed by Kufri Jyoti-ARC (3.1) compared to other treatments. Conclusively, ARC technology can be recommended for commercial cultivation and can ultimately be used for seed tuber production, timely supply at cheaper cost.

Keywords: *Potato, Apical Rooted Cuttings (ARC), Growth, Yield*

Introduction

Potato (*Solanum tuberosum* L.) is one of the most important tuberous, starchy and versatile food crop belongs to family Solanaceae. In India potato is grown in an area of 2.07 Mha with a production of 48 MT and a productivity of 23.19 MT/ha. The potato producing major states in India are Uttar Pradesh, West Bengal, Bihar, Madhya Pradesh, Gujarat and Karnataka. In Karnataka, the crop is grown in an area of 44,160 ha with an annual production of 5,89,120 MT and productivity of 13.34 t/ha (*Anonymous,* 2014). Further in Karnataka Potato is mainly grown in Hassan, Belgaum, Dharwad, Chikkaballapur and Kolar districts. Hassan district alone contributes more than 41 per cent of potato production in the state (*Bhajantri* 2011). Basically potato prefers relatively low temperature during early growth and cool weather during tuber development. It requires 25°C temperature at the time of sprouting, 20°C for vegetative growth and 17 to 20°C for tuber development. ARC saplings are mainly used for seed tuber production at a quicker rate. Farmers can produce zero generation seed tubers from ARC saplings without any shipping costs involved. In this regard KVK, Hassan after knowing the existence of innovative ARC sapling technology from the International Center for Potato (CIP), German innovation center (GIZ/GIC) Bengaluru, Krishi Vigyan Kendra, Kandali, Hassan procured tissue culture bottles/mother culture from CPRI, Shimla Himachala Pradesh through CIP, GIZ/GIC Bengaluru. By using the mother culture KVK Hassan has produced approximately two lakhs ARC saplings with existing facility and conducted the field evaluation studies on “Performance and evaluation of Apical Rooted Potato saplings in Hassan district”.

About the Technologies Addressed

The ICAR-KVK, Hassan has introduced two technologies in potato during last decade to improve the yield and to reduce the cost of production, namely, Apical Rooted Cuttings (ARC) for quality seed tuber production and Mechanization in Potato to reduce the labour requirement and drudgery.

The ARC is a technology where small portion of disease free mother tuber is collected and multiplied under aseptic conditions (Tissue culture) in a growing media to produce new and healthy plantlets. These plantlets are used as mother plants and transplanted to beds in nursery. The cuttings grown out of mother plants are replanted on the same beds. This process will continue upto 3 months where every mother plant provides 256 plants. The cuttings from these plants are planted in portrays and maintained upto 14 days' till rooting initiates and then transplanted to main field for seed production.

Methodology followed for commercial seed tuber production from ARC saplings

Disease free mother cultures (CPRI)

↓

In-vitro Multiplication in TC lab

↓

Mother plants multiplication in polyhouse

↓

Multiplication of ARC Seedlings in polyhouse

↓

Seed production (Season 1) in open field

↓

Seed production (Season 2) in open field

↓

Commercial potato production

No. of evaluation studies: 20 Trials

Places of studies conducted

Hassan Taluk	Tejuru, Bogarahalli, Maranahalli, Honnavara, Rudradevarahalli, Kammarige, Devihalli, soppanahalli, Beekanahalli, Kattaya and Honnavara, KVK, Kandali, Hassan
Arakalagudu Taluk	Kanchenahalli and Shanubogarahalli Villages
Alur Taluk	Bageshpura Villages
Holenarsipura Taluk	Yalachaganahalli Villages

Extension Management Practices

Initiatives: Capacity building programs and educational activities by KVK Hassan

- Conducted Training programs (On & off campus) - 16 No's
- Conducted Exposure visits to the farmers - 21 No's (483 farmers)
- Conducted Field visits - 64 No's
- Conducted Field days - 09 No's (509 farmers)
- Conducted Group discussion - 13 No's
- Conducted Radio Programme - 03 No's
- Conducted Television Programme - 01 No's
- Conducted Publications - 11 No's

Reach: KVK Hassan (UAS), Bengaluru is the first institute in the country who has taken up the evaluation studies and proved the suitability of ARC saplings in Hassan with the association of other organisations, and also under the financial assistance of Government of Karnataka. KVK Hassan is involved in multiplying ARC saplings and selling to farmers at less rate to promote own seed potato tuber production to reduce the dependability on seed sellers for potato seed tubers. KVK Hassan also involved in promoting private nurseries for taking up ARC saplings production to increase the availability of saplings to farmers at a quicker rate at cheaper cost. Conclusively, ARC technology can be recommended for commercial cultivation and can ultimately be used for seed tuber production which helps in timely supply at a lesser cost and to help farmers to get higher net returns and Benefit cost ratio (Table 1).

Table 1. Major differences observed before and after intervention of KVK

Before Intervention	After Intervention
Cut tuber/whole tuber planting	Apical rooted cuttings (ARC) saplings transplanting
Poor /uneven germination	Good establishment
Seed treatment is required	Not required
Chances of tuber rotage planting time	Less mortality and good seedlings establishment
Less vegetative growth and less yield	Good vegetative growth and good yield
Chances of spreading diseases through tubers	Diseases free planting materials
Use of old, low yielding varieties due to shortage of seed tubers (Ex. K.Jyothi)	Multiplication of good yielding varieties and supply of ARC is easy and fast (Ex.K.Himalini)
Seed cost is high	ARC cost is Less
Transportation cost is high	Local production of ARC
Less employment generation in commercial potato production	Good employment generation through ARC nursery, seed tuber production
Availability of 6 to 7th generation seed tubers in the market	Farmers can produce zero generation tubers from ARC
Farmers depending on vendors for seed	Farmers can produce own seed
Cut tuber/whole tuber planting	Apical Rooted Cuttings (ARC) transplanting
Poor /uneven germination	Good establishment

Technology Impact: The introduction of this technology has increased the yield by 12.34 per cent (Table 2). During the last five years' technology helped in production of 42 lakh potato saplings in the district, and has provided employment to 29 agricultural workers. This technology has spread to 65 hectares, and produced seed tubers worth of Rs. 81.76 lakhs which reduced the procurement of seed tuber cost in the district.

Table 2. Establishment percentage in different genotypes of Potato

S. No.	Genotypes	Establishment (%) 20 DAP
1	Kufri Jyoti-ARC	97.79
2	Kufri Jyoti-Tubers	93.48
3	Kufri Himalini-ARC	97.88
4	Kufri Himalini-Tubers	93.69
5	Mean	95.71
6	SEm±	0.15
7	CD at 5%	0.43

*DAP – Days After Planting

Table 3. Growth performance of different genotypes of potato at various stages

S. No.	Genotypes	Plant height (cm)		Number of branches / plant	Number of leaves / plant
		45 DAP	60 DAP	60 DAP	60 DAP
1	Kufri Jyoti-ARC	53.62	63.73	4.59	64.33
2	Kufri Jyoti-Tubers	52.35	62.44	3.59	55.83
3	Kufri Himalini-ARC	55.70	90.49	5.80	82.03
4	Kufri Himalini-Tubers	53.31	82.47	4.74	72.67
5	Mean	53.75	74.78	4.68	68.71
6	SEm±	0.20	0.17	0.02	0.18
7	CDat 5%	0.57	0.49	0.07	0.53

*DAP – Days After Planting

Table 4. Yield performance of different genotypes of potato

S. No	Genotypes	No. of tubers/ plant	Yield/plant (g)	Total yield (t/ha)
1	Kufri Jyoti-ARC	5.27	242.53	17.45
2	Kufri Jyoti-Tubers	4.32	216.11	16.46
3	Kufri Himalini-ARC	8.58	252.74	19.53
4	Kufri Himalini-Tubers	7.50	242.94	19.24
5	Mean	6.42	238.58	18.17
6	SEm±	0.06	0.35	0.08
7	CDat 5%	0.19	1.00	0.25

Apicals transplanting Seedlings at 12th day Seeglings ready for planting

Results: Results indicated that among the four treatments, Kufri Himalini-ARC(T3) recorded significantly higher percent of establishment (97.88 %), maximum plant height (90.49 cm), higher no. of branches (5.80), no. of leaves (82.03), no. of tubers per plant (8.58), total tuber yield per plant (252.74 g) and total tuber yield of 19.53t/ha followed by Kufri Jyoti-ARC (T1) with 97.79 % of establishment, plant height (63.73 cm), no. of branches (4.59), no. of leaves (64.33), no. of tubers per plant (5.27), total tuber yield per plant (242.53 g) and total tuber yield of 17.45 t/ha. The lower percent of establishment (93.48%), lower plant height (62.44 cm), no. of branches (3.59), no. of leaves (55.83), no. of tubers per plant (4.32), total tuber yield per plant (216.11 g) and total tuber yield of 16.46 t/ha was recorded by Kufri Jyoti-tubers (T2) (Tables 3 and 4). This might be due to their inherent genetic makeup, response to environmental condition, adaptability to particular environment and promotion of growth of auxiliary buds into new shoots (*Nikmatullah,* 2018).

Table 5. Cost of cultivation for Potato production in Hassan district

S. No.	Genotypes	Total cost of cultivation (Rs./ha)	Total yield (t/ha)	Gross returns (Rs./ha)	Net returns (Rs./ha)	B : C
1	Kufri Jyoti-ARC	1,37,500	17.45	4,36,250	2,98,750	3.1
2	Kufri Jyoti-Tubers	1,47,500	16.46	1,81,060	33,560	1.2
3	Kufri Himalini-ARC	1,37,500	19.53	4,88,250	3,50,750	3.5
4	Kufri Himalini-Tubers	1,47,500	19.24	2,11,640	64,140	1.4

Conclusion

In the present investigation it could be concluded that, among the different treatments, genotype Kufri Himalini-ARC, Kufri Jyoti-ARC and Kufri Himalini–tubers were found promising interms of total tuber yield, net returns and B:C ratio (Table 5) in the Hassan district during kharif season. Hence, ARC are suitable for cultivation for seed production in Hassan district.

References*

1. Anonymous. 2014. National Horticulture Board, Statistical data.
2. Bhajantri, S. 2011. Production processing and marketing of Potato in Karnataka an economic analysis. MBA (Agri.) thesis. University of Agricultural Sciences, Bengaluru, Karnataka, India, *pp 117.*
3. Lutaladio, N. B., Castaldi, L. 2009. Potato the Hidden treasure. *J Food Comp Annal.*, 22: 491-493.
4. Nikmatullah, A., Ramadhan, I., Sarjan, M. 2018. Growth and yield of apical stem cuttings of white Potato (*Solanum tuberosum* L.) derived from disease – free G-0 plants. *J. Appl. Hort.*, 20 (2): 139-145.
5. Rajput, A. M., Verma, A. R., Jain, S.K. 2003. Relative profitability of Potato varieties in Indore district of Madhya Pradesh. *Annals Agric. Res.,* 24 (2): 437-439.

* The content of this chapter was prepared in consultation with above references.

12

The Friends of Coconut Tree (FOCT): An Innovative Training Approach for Self-Employment of Farm Youth in Tumakuru District

V. Govinda Gowda[1], Y.N. Shivalingaiah[2], M.E. Darshan[3] and D. Harshitha[4]

[1]Sr. Scientist & Head, ICAR-KVK, Konehally, [2]Director of Extension, [3]Scientist (Agril. Extension), ICAR-KVK, Konehally, [4]Assistant Professor (Agril. Extension), Directorate of Extension, UAS, Bengaluru

Email: drharshitha@gmail.com

Abstract

The Friends of Coconut Tree (FOCT) training program, initiated by ICAR-KVK Tumakuru-I in collaboration with the Coconut Development Board (CDB) has successfully addressed the shortage of skilled labour in coconut harvesting as a Para Extension Worker Since 2012 and trained 352 rural youth in mechanized climbing techniques leading to a 34.85 per cent self-employment rate and an average monthly income of Rs. 25,000-30,000. Participants efficiently harvest 40–50 trees daily by charging Rs.50 per tree which contributing to rural economic stability. Group-based entrepreneurship such as the formation of Kalpasiri and Kalpaganga Coconut Climbers Groups has further enhanced labor efficiency and collective income generation in rural areas. This initiative mitigates urban migration, fosters entrepreneurial ventures and ensures a sustainable and skilled workforce for coconut industry in Karnataka. Scaling efforts include expanding training programs, subsidizing mechanized tools, forming cooperatives, leveraging digital job-matching platforms and fostering public-private partnerships. The FOCT program exemplifies agricultural innovation, skill-based economic mobility and rural livelihood enhancement by assuring employment opportunities to farm youths in Tumakuru district of Karnataka state.

Keywords: *FOCT (Friends of Coconut Tree) Training Program, Mechanized Climbing Techniques, Self-Employment, Entrepreneurship, Rural Economic Stability*

Introduction

Nestled in the heart of Karnataka, Tumakuru District is renowned for its verdant landscapes dotted with vast expanses of coconut and areca nut plantations. The region's unique climatic conditions, characterized by a harmonious blend of sunshine and seasonal rains, create an ideal environment for the growth of coconut palms, which have become a staple of both the local economy and the cultural fabric of the community.

Historically, coconut trees have not only provided sustenance but have also played a pivotal role in the livelihoods of many families in Tumakuru. The fruit is a key ingredient in various local dishes and is widely utilized in traditional ceremonies and celebrations. However, the task of harvesting coconuts from these towering trees has long been a physically demanding endeavour. Traditionally, skilled climbers would ascend the tall, slender trunks, utilizing their agility and experience to pluck coconuts by hand. This craft, which once flourished, has now seen a decline, making it increasingly challenging for coconut planters to find professionals who can perform this labour-intensive task.

The decline in manpower has made it evident that traditional methods are no longer sustainable. If coconut farmers wish to keep pace with growing market demands, a new approach is necessary one that balances traditional harvesting practices with modern technology. Recognizing this need, innovative minds have introduced the "Coconut Tree Climber", a mechanized device specifically designed to address the challenges faced by coconut planters.

The Coconut Tree Climber is a user-friendly and efficient device that significantly simplifies the process of climbing and harvesting. Built with a strong and adjustable mechanism, the device allows a person to securely climb tall coconut trees with minimal physical effort. More importantly, it drastically reduces the risk of accidents, ensuring the safety of the user. Unlike traditional methods that rely solely on human strength, the device takes advantage of mechanical power, enabling planters to harvest coconuts faster, safer, and more efficiently.

This innovation has proven to be a game-changer in the coconut farming industry. Not only does it empower farmers to become self-reliant, but it also minimizes their dependence on skilled climbers. Additionally, the device can be operated by individuals of varying ages and physical strength, making it accessible to a broader range of people. This, in turn, provides new employment opportunities for those who may not have had the physical ability or training to climb trees in the past.

Another significant advantage of the Coconut Tree Climber is its ability to meet growing market demands. With faster and more efficient harvesting, planters can now ensure a steady supply of coconuts, preventing potential losses caused by delayed harvesting. This increased productivity not only benefits the planters but also strengthens the coconut industry as a whole, allowing it to remain competitive in the global market.

Moreover, the introduction of mechanized tree climbers also encourages young entrepreneurs and rural workers to stay connected to agriculture. By integrating technology with traditional farming practices, the perception of agricultural work is slowly being redefined, attracting more people to engage in farming activities without hesitation.

In conclusion, the Coconut Tree Climber stands as a perfect example of how innovation can transform age-old practices, bringing convenience, safety, and efficiency to the forefront. It not only helps overcome labor shortages but also preserves the essence of coconut farming while aligning it with the demands of the modern world. As technology continues to advance, such devices are paving the way for sustainable and prosperous agriculture, ensuring that the art of coconut harvesting continues to thrive for generations to come.

Technologies Addressed

To address the ongoing challenges posed by the dwindling availability of skilled coconut tree climbers, ICAR KrishiVigyan Kendra (KVK) Tumakuru-I has taken a proactive step by introducing a vocational training program titled "Friends of Coconut Tree (FOCT)". This initiative was launched in collaboration with the Coconut Development Board, Bengaluru, aims to equip rural youth with specialized skills in mechanized coconut tree climbing and harvesting. The program not only focuses on safe and efficient techniques for using modern coconut-climbing devices but also provides participants with comprehensive knowledge about coconut production practices—from sowing to harvesting. By offering hands-on training, the program ensures that young individuals gain both technical proficiency and confidence in managing various aspects of coconut farming.

The FOCT training program goes beyond just skill development; it serves as a pathway for employment and self-reliance among rural youth. Participants are trained not only in harvesting techniques but also in crucial aspects of coconut cultivation, including nursery management, planting methods, soil fertility management, irrigation practices, pest and disease control, nutrient management, and post-harvest processing. This holistic approach helps them develop a deep understanding of coconut farming as a sustainable livelihood option. With the growing demand for coconut harvesting services, trained individuals can either take up independent contracts or form service groups to cater to coconut plantations across the region. Additionally, the program fosters entrepreneurship by encouraging trainees to establish their own coconut harvesting and management enterprises, thereby generating sustainable livelihood opportunities.

By bridging the gap between traditional agricultural practices and modern technological advancements, this initiative not only revitalizes the coconut industry but also plays a pivotal role in rural economic development. It empowers young individuals to see coconut farming as a profitable and viable career path, thereby preventing migration to urban areas in search of jobs. Ultimately, the FOCT program serves as a crucial step toward strengthening the coconut-based economy while preserving the agricultural heritage of Tumkur and beyond

Extension Management Practices in the FOCT Training Program

1. Initiatives

ICAR KVK Tumkur-I, in collaboration with the Coconut Development Board (CDB), Kochi, Kerala, has proactively implemented the FOCT training program to address the labor shortage in coconut harvesting. This initiative was launched to equip rural youth with mechanized tree-climbing skills and knowledge of coconut production practices, plant protection, and post-harvest management. The program focuses on skill development, employment generation, and entrepreneurship promotion, ensuring a sustainable livelihood for participants. Additionally, KVK has introduced group-based approaches to encourage teamwork and problem-solving, forming Kalpasiri and Kalpaganga Coconut Climbers Groups.

2. Reach

Since its inception in 2012, the program has been progressively expanded, covering:

- 350 young farmers since 2012 under the palm climbing and plant protection training program.
- Continuous engagement of around 70 trained individuals actively working in coconut harvesting, crown cleaning, and plant protection activities.
- Formation of two dedicated climbers' groups:
  - Kalpasiri Coconut Climbers Group (Raysandra, ThuruvekereTaluk)
  - Kalpaganga Coconut Climbers Group (Ganganahalli, Thuruvekere Taluk)

- Trained individuals have also been invited as Master Trainers in other KVKs to impart their skills, further expanding the outreach of the program.
- The trained climbers have harvested coconuts across South India, covering a total of 73,000 hectares.
- This climbing machine technology has reached 1,25,000 farmers across South India through the efforts of trained climbers
- The trained climber has imparted palm climbing skills to 65 young individuals.

3. Output

The FOCT training program has resulted in significant practical outcomes, including:

- Increased participation in self-employment:34.85 per cent of trained youths are now self-employed as professional coconut climbers.
- Economic benefits:
  - A single climber can harvest 40-50 coconut trees per day.
  - Climbers charge Rs.50 per palm for harvesting and Rs.50 for crown cleaning and plant protection and earn Rs 2500 -3000 per day
  - Monthly earnings range between Rs. 25,000 -Rs. 30,000 per person.
- Labor shortage mitigation: Farmers can now access skilled workers for timely harvesting and plant maintenance.
- Encouragement of group-based entrepreneurship: The formation of climbers' groups enables efficient labor management and supports cooperative income generation.
- Trained climbers have collectively generated a revenue of 85 lakhs through palm climbing entrepreneurship.

4. Impact

The FOCT training program has positively transformed the coconut farming sector by:

- Enhancing rural employment opportunities and encouraging youth engagement in agriculture.
- Bridging the labor gap in coconut harvesting, which was previously dependent on traditional climbers.
- Providing sustainable income sources, reducing migration to urban areas for employment.

- Empowering rural youth to become entrepreneurs, with many now running independent coconut harvesting services.
- Expanding horizontal learning opportunities, where trained climbers are now trainers at other KVK programs, spreading knowledge and creating a multiplier effect further.
- Trained climbers provide crown cleaning and pest control services, leading to a 25per cent increase in coconut yield compared to previous levels.

Overall, KVK Tumkur-I's extension management practices in FOCT training have strengthened the coconut industry, improved economic stability for rural youth and enhanced the efficiency of coconut harvesting and plant protection practices in the region.

The Inspiring Journey of Sri. Devananda – From Struggles to Success

Sri. Devananda, the son of Late Rammayya, hails from Chickabidare village. He is a marginal farmer with a small landholding of 3 acres of dry land. In order to sustain his livelihood, he took up multiple jobs—working as a farm laborer, a lorry driver, and a farmer. His daily earnings ranged between Rs. 200-250, making it difficult to achieve financial stability. Despite the hardships, his determination to improve his economic condition kept him moving forward.

The Turning Point – Intervention

Sri. Devananda came across an advertisement from the Krishi Vigyan Kendra (KVK), Tiptur, about a training program called "Friends of Coconut Tree", sponsored by CDB, recognized an opportunity to learn a new skill, he eagerly enrolled in the program.

The one-week training program covered various aspects of palm climbing, harvesting techniques and plant protection technologies. He has actively participated in all sessions, mastering the art of coconut tree climbing, crown cleaning and pest management. This training proved to be a life-changing experience for the farmer.

Impact – The Road to Prosperity

Armed with his newfound skills, Sri. Devananda began harvesting coconuts from his own farm. Later he has extended his services to neighboring farms, charging Rs. 50 per palm for harvesting. His expertise and dedication quickly made him a highly sought-after professional in his district and even in neighboring districts.

With an increased demand for his services, Sri. Devananda's income grew significantly. He now earns between Rs. 2000 - Rs. 2500 per day, amounting to a minimum of Rs. 25,000 per month from coconut climbing alone, in addition to his daily farming activities.

Recognizing the growing need for skilled coconut climbers, he began mentoring other trainees from the KVK training programs. Whenever demand surged, he collaborated with fellow trained individuals to complete harvesting tasks efficiently.

His expertise and dedication earned him the role of Master Trainer in FOCT training programs organized by KVK and other agricultural organizations. Encouraged by his success, Sri. Devananda diversified his income by starting an animal husbandry enterprise, where he rears sheep and goats alongside his farming activities.

Spread Effect & Community Development

Sri. Devananda's success story has inspired many in his village. Seeing the positive impact of the training, 10 young farmers from Chickabidare registered for the FOCT training at KVK, Tiptur. They learned valuable skills in palm climbing and plant protection, following in his footsteps.

Moreover, his family members purchased a coconut tree climbing machine and are now practicing under his guidance. His journey from a struggling farmer to a financially stable and well-respected trainer has set an example for the entire farming community.

Conclusion

The entrepreneur proudly acknowledges the training program for transforming his life. With dedication, hard work and the right skills, he has not only improved his own economic status but has also contributed to the development of his village and fellow farmers. His journey stands as a testament to the power of skill development and uplifting rural livelihoods. Today, is not just a farmer; he is a mentor, a role model, and a true "Friend of Coconut Trees."

Conclusion and Strategic Implementation Plan

The FOCT training program spearheaded by ICAR-KVK Tumakuru-I have proven to be a transformative initiative in addressing the skilled labor shortage in coconut harvesting while simultaneously promoting employment and entrepreneurship among rural youth. By integrating modern mechanized climbing techniques with traditional agricultural practices, the program has revitalized the coconut industry, creating sustainable livelihoods and reducing urban migration.

To ensure the long-term success and scalability of this initiative, a strategic implementation plan is essential:

1. Scaling Training Programs – Expanding the FOCT training program across more districts to train a larger number of youth, ensuring wider access to employment opportunities.
2. Technology Dissemination – Promoting the widespread adoption of mechanized coconut tree climbers by subsidizing equipment costs and offering financial assistance through government schemes.
3. Entrepreneuria Development – Establishing Coconut Climber Cooperatives and Self-Help Groups (SHGs) to facilitate collective service models, increasing efficiency and bargaining power for climbers.
4. Public-Private Partnerships – Collaborating with agribusiness companies, research institutions, and government agencies to enhance training quality and market linkages.
5. Digital Integration and Awareness – Leveraging mobile applications and online platforms for job matching between trained climbers and coconut farmers, improving accessibility and service efficiency.
6. Sustainability & Monitoring – Implementing regular follow-ups, refresher courses, and performance tracking to ensure ongoing skill enhancement and adaptation to emerging agricultural technologies.

By strategically implementing these measures, the FOCT initiative can significantly contribute to the economic empowerment of rural youth, ensuring a sustainable, skilled workforce for the coconut industry while fostering agricultural resilience and rural development in Karnataka and beyond.

References*

1. Anonymous. 2023. Annual Report on FOCT Training Programme, ICAR - KVK, Tumakuru-I, University of Agricultural Sciences, Bengaluru
2. Kumar, P., and Reddy, S. 2022. Mechanized Coconut Climbing and Harvesting: A Solution to Labor Shortage. *J. of Agril Engi*, 59(3), 89-97.
3. Raghavendra, S. 2023. Impact of Mechanized Harvesting on Coconut Production. *Agricultural Review*, 45(4), 120-130.
4. Sujatha, R. 2021. Rural Youth and Agriculture: Challenges and Opportunities. *Economic and Political Weekly*, 56(19), 78-85.

* The content of this chapter was prepared in consultation with above references.

13

Impact of Plant Health Clinic Activities on Major Agriculture and Horticulture Crops of Chamarajanagara District

G.S. Yogesh[1], Y.N. Shivalingaiah[2], B. Pompanagouda[3], Ashoka Doddamani[4] and V. Bharat[5]

[1]Senior Scientist & Head, ICAR-KVK, Chamarajanagara, [2]Director of Extension, [3]Senior Technical Officer, ICAR-KVK, Chamarajanagara, [4]Scientific Officer, [5]Scientist (Agricultural Extension), ICAR-KVK, Chamarajanagara, UAS, Bengaluru.

Email: deuasBengaluru@gmail.com

Abstract

Chamarajanagara is the southernmost district in the state of Karnataka. Total areas under agricultural and horticultural crops are about 193657 ha and total production 615910 tonnes. The major crops viz., Jowar, Maize, Paddy, Ragi, Horsegram, Cowpea, Greengram, Sunflower, Groundnut Banana, Turmeric, Coconut, Arecanut, Tomato, Chili and Brinjal. The average yield losses in the major crops in the district are about 15-20 per cent due to insect pest and diseases, the major insect pest and diseases viz., Fall armyworm, Thrips, Mites, Diamondback moth, Tuta, Fusarium wilt, Late blight, Early blight. Sigatoka leaf spot, Twister disease etc., to combat these problems KVK has intervened at district level by establishment of Plant Health Clinic. where in the plant health clinic first diagnosed problem and based on the recommendation (PHC) will be suggested. This suggestion will be majorly on bio-intensive-based and will provide quality bio-based products to the farmers. From last three years, PHC has provided 25443 kg of bio-pesticides viz., Trichoderma, Pseudomonas, Paeciliomyces, Metarhizum and Beauveria a total of 5,125 farmers were benefitted. As a result, 10 -15 per cent yield increased in major crops and additional income of Rs.10000 per acre increased per farmer. Further a worth of Rs. 37,45,825 pesticide consumption has been reduced over three years.

Keywords: *Agriculture, Bio-pesticides, Horticulture, Plant Health Clinic*

Introduction

Globally, the agricultural sector plays an important role in poverty reduction, food security and promotion of nutritious food (Jones & Ejeta, 2016). Although agricultural crop production is essential, there are challenges associated with plants due to pests and diseases infestation in the value chain of crop production. Plant health problems can be resolved through appropriate diagnosis and identification of plant pests affecting agricultural (Ausher *et al.,* 1996). However, Oerke (2006) reported that 40 per cent of crop yield losses are caused by plant pests and diseases of economic importance. The majority of smallholder farmers are faced with risks associated with pests and diseases in plant production.

To address challenges faced by smallholder farmers, plant health clinics are important for plant pests and disease diagnosis. The plant health clinics can be defined as a system in which farmers obtain advisory services on the plant pest problem affecting agricultural crops and provide management control strategy for pest management (Alokit *et al.,* 2014). There are many plant health clinics initiatives within the Indian region that provides plant pests and diseases advisory services to all categories of the farmers (Boa *et al*., 2016). Many plant clinics in developing countries which are run by international research institutions are more efficient and effective as compared to those not run by the research institutions. This review seeks to investigate and assess the state of global plant clinics with the aim of obtaining the efficient and effective plant health clinic model for South Africa. The knowledge, awareness, availability and accessibility of basic plant healthcare to smallholder farmers for plant pests and diseases diagnosis to ensure early detection, pest management and pest reporting remain a challenge. Furthermore, there is limited literature and research on plant health clinics' establishment and its importance to smallholder farmers in the Indian developing country.

Background of the Plant Health Clinics

To date, there are about 34 countries operating plant health clinics globally and it is estimated that there are 4,500 plant clinics across the globe Centre for Agriculture and Biosciences International (CABI), 2020; Tambo *et al.,* 2020). These include plant clinics in Africa, South America and Asia respectively (Boa, 2009; CABI 2020). The first plant health clinics were established in 2003 in Bolivia (Bentley *et al*., 2009; Boa, 2009). In Africa alone, plant clinics were established with the help of CABI to assist smallholders and household to respond effectively and efficiently towards plant pests and diseases. According to Boa *et al.* (2016), there are two types of plant health clinics in terms of service provisions, namely (1) Fixed structure with appropriate laboratory equipment for identification of pests and diseases (2) extension support through mobile plant clinic or community-based clinic. Mobile plant health clinic is the most valuable tool for smallholder farmers' accessibility. According to Adhikari *et al.* (2017), plant health clinic is a critical system approach to provide immediate service to smallholder farmers. Although plant clinics are critical, Majuga *at el.* (2018) indicated that not much is known on the operation of the clinics and their importance. Accordingly, plant health clinics can prompt official pest reporting obligations, policy formulation, official pest status of a country, as well as official early warning and/or pest management actions (FAO, 2011, RSA, 2017, DAFF, 2017). In the year between 2000-2009, plant health clinics in Bolivia were found to assist large number of communities and farmers in terms of provision of advisory services on disease and pestswhich led to less usage of pesticides (Bentley *et al.,* 2011).

Globally, the impact of plant health clinics is to diagnose pests and diseases timely and to ensure food production and security thereby employing integrated pest management strategy (Srivastava, 2013). The Republic of South Africa (2017) outlines the importance and procedure for pest reporting procedures of regulated pests and/or new pest. Pest reporting promotes sharing of data and knowledge (Plantwise, 2014), Boa (2015) indicated in countries where plant clinics exist, awareness to inform farmers to bring diseases plant samples for appropriate advice and recommendations to manage and control the disease is critical. The Food and Agriculture Organisation (2017) described plant pest surveillance as an important element for pest reporting obligation within trading partners. In the United State of America, surveillance to detect a new pest or pest of economic importance is conducted through the Cooperative Agricultural Pest Survey Program (CAPSP) which includes industry, farmers and universities. This program does not only consist of government authority (Jackson, 2011).

Plant health clinics are beneficial for smallholder farmers and rural communities by means of enhancing crop productivity and food security. Van der Linde and Rong (2019) provided fundamental basis of establishing of plant clinic to benefit smallholder farmers in South Africa The study conducted by Bentley *et al.* (2011) revealed that plant clinics could have a huge impact in terms of farmers' income return especially for those receiving trainings. The other positive impact for plant clinics identified includes providing advisory on plant health problems and gaps, recommendations on the safe use of pesticides and pest management and control strategies (Boa *et al.,* 2016). Adhikari *et al.* (2015) listed some of the impact outcomes of plant clinics which include: knowledge and awareness, increased knowledge and skill on plant health problems on extension support staff and increased extension services to the smallholder farmers.

Model for a sustainable, efficient and effective use of plant health clinics The model for Plant clinic can be defined as a mechanism in which the smallholder farmers obtained primary plant health advisory services relating to various plant health problems on their crops. The model differs from country to country, however, Centre for Agriculture and Bioscience (CABI) preferred model wherein there are collaborations between spheres of government: National, Provincial and Local government.

In other countries, there are various systems in which funding is coordinated, for example, sponsorship, government as well as state own enterprises (Jenner, 2019). The model for plant clinics can be similar to those of animal healthcare (Bentley *et al.*, 2009). Plant clinics differ: there are fixed and mobile clinics (Boa *et al.*, 2016; Kelly, 2009). Appropriate model for plant health clinic to enhance services delivery in terms of coverage, access and the quality of plant healthcare is found to be a mobile system. In assessing various plant health clinics approaches, Jenner (2019) provided some of the key functional areas of plant health clinics as follows: • Linkages among stakeholders under plant health system, • Inclusive approach to all relevant stakeholders, • Increase interaction between extension officers and farmers, • Provide a diagnosis services and solution to a plant health problem, • Provide early detection of pest and reporting to the relevant authority, • Record keeping of plant pest problem and other information.

Plant Health Clinic at ICAR KVK Chamarajanagara

The major crops are grown in Chamarajanagar district are jowar, maize, paddy, ragi, greengram, blackgram, horsegram, cowpea, sunflower, groundnut, banana, turmeric, coconut, arecanut, tomato, chilli, brinjal and watermelon. The average yield losses in the major crops in the district are about 15-20 per cent due to insect pest and diseases i.e. fall armyworm, thrips, mites, diamondback moth, pinworm (*Tuta*), *Fusarium* wilt, rhizome rot, late blight, early blight, sigatoka leaf spot, twister disease, etc. To address these issues, KVK has intervened at district level by establishing Plant Health Clinic (PHC) on 27.08.2011 and it was inaugurated by Dr. S. Ayyappan, Secretary, DARE and Director General, Indian Council of Agricultural Research, New Delhi

Activities Carried Out Through Plant Health Clinic

- **Diagnosis and recommendation:** diseased/infected/infested plant samples bought by farmers are diagnosed and recommended, scientific and specific management practices
- **Capacity building of rural youth:** rural youths trained on production and use of bio-pesticides *viz., Trichoderma, Pseudomonas, Paecilomyces, Metarhizium and Beauveria*
- **Other activities:** identification of crop pests & diseases, integrated pest & disease management and safe use of pesticides organizing plant health awareness camps, disseminating plant protection advisories through print media and electronic media
- **Production and sales of bio-pesticides to the farmers**: *Trichoderma harzianum, Trichoderma viride, Pseudomonas fluorescens, Paecilomyces lilacinus, Metarhizium anisopliae and Beauveria bassiana*

Trichoderma

Pseudomonas

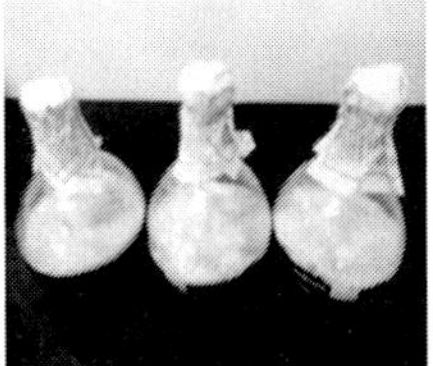

Paeciliomyces

Metarhizium

Output and Outcome

Since 2021 to till date, 5125 farmers of the district have visited the clinic, taken advisories & recommendations and have been benefitted with respect to insect-pests and disease management in agriculture & horticulture crops

Table-1: Production and sales of bio-pesticides (Trichoderma, Pseudomonas, Paecilomyces, and Metarhizium) from 2020-21 to 2022-23 at KVK PHC

Year	Production (l/Kg)	Gross Incom (□)	Expenditure (□)	Net Income (□)	No of farmers benefitted
2020-21	4,828	4,82,881	1,20,572	3,62,309	965
2021-22	10,884	10,88,400	1,44,686	9,00114	2176
2022-23	9,721	9,72,100	2,84,909	7,07,201	1984
Total	**25,443**	**25,43,381**	**5,50,167**	**19,69,624**	**5125**

- More than 57 per cent of the farmers repeatedly visited PHC for procuring bio-pesticides successively
- Total 25,443 l/kg of bio-pesticides *viz., Trichoderma, Pseudomonas, Paecilomyces*, *Metarhizium* and *Beauveria* have been provided
- More than 10 -15 per cent increase in yield of major crops has been recorded with an additional income of Rs.10,000 per acre as a result of reduction in number of plant protection chemicals sprays by 2 to 3 per acre in major crops of the district
- In total, a worth of Rs.37,45,825 pesticide consumption has been reduced over three years
- Net income of Rs. 19,93,214 earned over three years (2021 to 2023)

Fig 1. Enrichment of manures with bio-agents and its application to horticultural crops

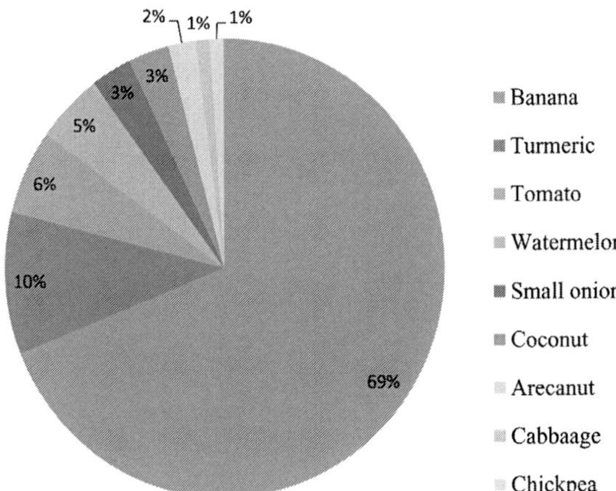

Fig 2. Percentage utilization of bio-control agents in different agricultural and horticulture crops in Chamarajangara district (2021-20 to 2023-24)

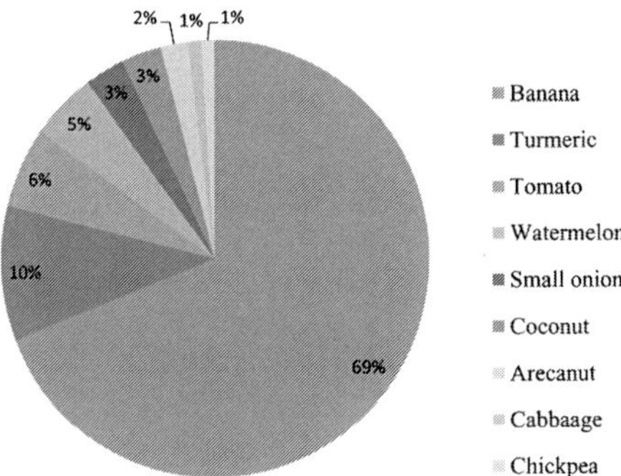

Fig 3. Percentage utilization of bio-control agents in different agricultural and horticulture crops in Chamarajangara district (2021-24)

Conclusion

Through Plant Health Clinic, 5125 farmer of the district have been benefitted through recommendations pertaining to management of major insect-pests and diseases such as fall armyworm, thrips, mites, diamondback moth, pinworm (*Tuta*), fusarium wilt, rhizome rot, late blight, early blight, sigatoka leaf spot, twister disease. Major bio-pesticides that have been procured by farmers of the district are *Trichoderma, Pseudomonas, Paecilomyces* and *Metarhizium* for managing insect-pests and diseases in crops like Banana, Turmeric, Tomato, Watermelon, Small Onion and Coconut. This has resulted in 10 -15 per cent increase in yield and has provided an additional income of Rs.8000 to Rs. 10,000 per acre.

Total, a worth of Rs.37,45,825 pesticide consumption has been reduced due to reduction in number of plant protection chemicals spray by two to three per acre and has benefited the ecosystem services.

References

1. Adhikari, D., Sharma, D. R., Pandit, V., Schaffner, U., Jenner, W. & Dougoud, J. 2017. Coverage and access of plant clinic in *Nepal. J. Agric. Environ.,* 18(2): 51-58.
2. Adhikari, R. K., Regmi, P. P., Thapa, R. B. & Boa, E. 2015. SWOT analysis of Plant Health Clinics as perceived by plant doctors in Nepal. *J. Inst. Agric. Anim. Sci.,* 19(6): 137-146.
3. Alokit, C., Tukahirwa, B., Oruka, D., Okotel, M., Bukenya, C. & Mulema, J. 2014. Reaching out to farmers with plant health clinics in Uganda. *Uganda J.Agric. Sci.,* 15(1): 15-26.
4. Arun Balamatti. 2019. Demand-driven Extension – The Case of JSS Agri Clinic by KVK, Mysuru, eSARD 2019 International Conference on Extension for Strengthening Agricultural Research and Development.
5. Ausher, R., Ben-Ze'ev, I. S. & Black, R. 1996. The role of plant clinics in plant disease diagnosis and education in developing countries. Annual review of phytopathology, 34(1) : 51-66.
6. Bentley, J. W., Boa, E., Danielsen, S., Franco, P., Antezana, O., Villarroel, B. & Herbas, J. 2009. Plant health clinics in Bolivia 2000—2009: operations and preliminary results. *Food Secur.,* 1(3): 371-386.
7. Bentley, J., Boa, E., Almendras, F., Franco, P., Antezana, O., Díaz, O. & Villarroel, J. 2011. How farmers benefit from plant clinics: an impact study in Bolivia. *International J. Agric. Sustain.,* 9(3) : 393-408.
8. Boa, E. and Harling, R. 2008. Starting plant health clinics in Nepal. *Global Plant Clinic-CABI, UK.*
9. Boa, E. 2009. How the global plant clinic began? Outlooks on Pest Management, *J. Integrative Agric.,* 20 (3): 112 116.
10. Boa, E., Franco, J., Chaudhury, M., Simbalaya, P. & Van Der Linde, E. 2016. Plant health clinics. What Works in Rural Advisory Services. GFRAS, Switzerland.
11. Cabi. 2015. Plantwise strategy 2015–2020, Wallingford, UK: CABI. Available at: http://www.plantwise.org/about-plantwise/strategy.
12. Danielsen, s., Boa, E., Mafabi, M., Mutebi, E., Reeder, R., Kabeere, F. and Karyeija, R. 2013. Using plant clinic registers to assess the quality of diagnoses and advice given to farmers: a case study from Uganda. *J. Agric. Edu. Ext.*, 19 (2): 183-201.

13 Department of Agriculture, Forestry and Fisheries (DAFF). 2017 Pest alert: Detection of a new pest caterpillar for the first time in South Africa. Media release.

14 Flood, J. 2010. The importance of plant health to food security. *Food Secur.,* 2(3): 215 231.

15. Food and Agriculture Organisation (FAO). 2017. International Standards for Phytosanitary Measures (ISPM 17). Pest Reporting. International Plant Protection Convention (IPPC). Rome.
16. Jackson, L. D. & Fieselmann, D. A. 2011. The Cooperative Agricultural Pest Survey (CAPS) program. Scientific support to optimize a national program. United States

Department of Agriculture: Animal and Plant Health Inspection Service (USDA: APHIS), Plant Protection and Quarantine (PPQ).
17. Jenner, W. 2019. Plant clinics as embedded services CABI Switzerland. Available from www.plantwise.org [Accessed 23 March 2019].
18. Jones, A. D. and Ejeta, G. 2016. A new global agenda for nutrition and health: the importance of agriculture and food systems. Bulletin of the World Health Organization, 94 (3): 228.
19. Majuga, J. C. N., Uzayisenga, B., Kalisa, J. P., Almekinders, C. & Danielsen, S. 2018. "Here we give advice for free": the functioning of plant clinics in Rwanda. *J. Econ. Dev.*, 28 (7): 858-871.
20. Oerke, E. C. 2006. Crop losses to pests. *J. Agric. Sci.,* 144(1): 31-43.
21. Plantwise. 2014. Plant Clinic data management. IPPC National Reporting Obligations and Plantwise. Nairobi, Kenya. www.plantwise.org.
22. Republic of South Africa (RSA). 2017. Agricultural Pests Act No.36 of 1983 (Act No.36 of 1983) as amended. Control Measures Amendment. R.1271 of 17 January 2017. Republic of South Africa. Pretoria.
23. Srivastava, M. P. 2013. Plant clinic towards plant health and food security. *International J. Phytopathol.*, 2 (3) :193-203.
24. Tambo, J. A., Uzayisenga, B., Mugambi, I. & Bundi, M. 2020. Do Plant Clinics Improve Household Food Security? Evidence from Rwanda. *J. Agric. Econ.,* 16 (6): 112-119.
25. Van Der Linde, E. & Rong, I. 2019. Mobile plant health clinics – a new approach to monitoring plant diseases in South Africa. Agricultural Research Council (ARC). Pretoria.

14

Effective Utilization of Social Media for Technology Dissemination by ICAR-KVK, Bengaluru Rural

***B.G. Hanumantharaya*[1]*, Y.N. Shivalingaiah*[2]*, Y.M. Gopala*[3] *and D. Harshitha*[4]**

[1]Sr. Scientist & Head, ICAR-KVK, [2]Director of Extension, [3]Scientist (Agril. Extension), ICAR-KVK, [4]Assistant Professor (Agril. Extension), Directorate of Extension, UAS, Bengaluru

Email: drharshitha@gmail.com

Abstract

ICAR-Krishi Vigyan Kendra (KVK), Bengaluru Rural has strategically embraced social media platforms to revolutionize agricultural technology dissemination. Recognizing the digital landscape's potential, the KVK utilizes the social medias YouTube, Facebook, Instagram, and X (formerly Twitter) to bridge the gap between research and farmers. This multi-platform approach ensures timely and accessible information delivery, fostering agricultural advancement. YouTube serves as a visual repository, hosting detailed video documentation of technologies released by the University of Agricultural Sciences (UAS), Bengaluru. These videos provide practical demonstrations and expert insights, simplifying complex concepts. Farmer testimonials on Frontline Demonstrations (FLDs) and success stories of progressive farmers offer authentic perspectives, inspiring adoption. Concise technology video capsules and recordings of regular activities KVK further enhanced knowledge transfer. Facebook facilitates community engagement through daily weather and market price updates, crucial for informed decision-making. The platform mirrors YouTube's video content, amplifying reach. Instagram leverages visual engagement with short video clips and images showcasing technologies and farmer achievements. X (Twitter) provides concise updates and links to detailed resources, ensuring rapid information dissemination. This strategic utilization of social media empowers farmers with readily available, relevant information, fostering the adoption of innovative agricultural practices. By leveraging the unique strengths of each platform, KVK Bengaluru rural effectively strengthens extension efforts and promotes sustainable agricultural development. The KVK's digital approach exemplifies a modern strategy for agricultural extension, demonstrating the transformative potential of social media in bridging the digital divide and driving agricultural progress.

Keywords: *Social Media, Extension, Frontline Demonstrations*

Introduction

Social media has turned into an essential element of individuals' lives students in today's world of communication. Its use is growing significantly more than ever before especially in the post – pandemic era, marked by a great revolution happening to the recent investigations of using social media show that approximately 3 billion individuals worldwide are now communicating via social media (Iwamoto and Chun, 2020). This growing population of social media users is spending more and more time on social network groupings, as facts and figures show that individuals spend 2 hours a day, on average, on a variety of social media applications, exchanging pictures and messages, updating status, tweeting, favouring and commenting on many updated socially shared information (Abbott, 2017).

In the digital age, social media have emerged as a powerful tool for disseminating agricultural technologies and knowledge. ICAR-Krishi Vigyan Kendra (KVK), Bengaluru Rural has effectively leveraged various platforms to bridge the gap between research and farmers, ensuring timely and relevant information reach the farming community. This article explores how KVK Bengaluru Rural strategically utilizes YouTube, Facebook, Instagram, and X (formerly Twitter) for technology dissemination.

I. Facebook

Facebook's interactive nature allows KVK Bengaluru Rural to engage with farmers and build a strong online community. The fact that KVK's Facebook channel has 5080 followers is a significant indicator of its reach and engagement within the farming community. The followers represent a substantial audience actively interested in the KVK's activities and information, this indicates that, the KVK has successfully established a relevant online community. With this follower base, the KVK can effectively disseminate critical information, such as weather updates, market prices, and technological advancements, to a large segment of the local farming population. A significant follower count suggests that the KVK is perceived as a reliable and trustworthy source of agricultural information, farmers are more likely to follow and engage with organizations they trust.

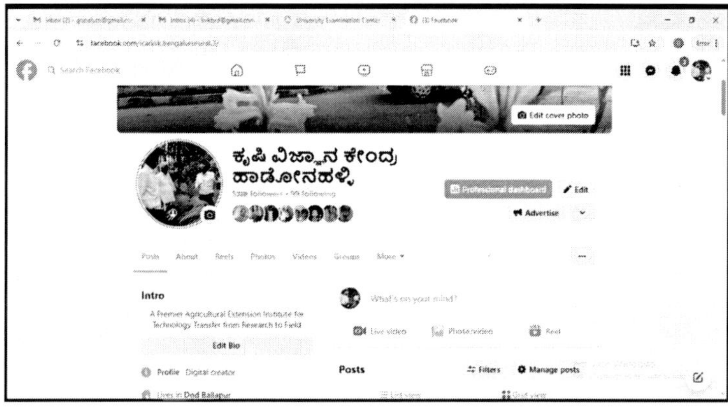

Fig 1. Home page of Facebook channel of ICAR-KVK, Bangaluru Rural

The Facebook channel serves as a valuable extension tool, complementing traditional methods. It allows the KVK to reach farmers who may not be able to attend physical workshops or events. The platform facilitates two-way communication, enabling farmers to ask questions, provide feedback, and share their experiences. The interaction helps the KVK understand the needs of the community and tailor its services accordingly. The KVK can effectively promote new agricultural technologies and best practices through its Facebook channel, encouraging adoption and improving agricultural productivity. In cases of sudden changes, such as unexpected weather events, or market fluctuations, the KVK can rapidly distribute information to a large audience.

While follower count is important, engagement (likes, comments, shares) is also crucial, Higher engagement indicates that followers are actively interacting with the content. Maintaining a high follower count requires consistently providing relevant and valuable content that meets the needs of the farming community. The KVK can further expand its reach by promoting its Facebook channel and engaging with other agricultural organizations and groups. In essence, the followers on ICAR-KVK, Bangaluru Rural's Facebook channel represent a valuable asset for the organization, enabling it to effectively fulfil its mission of disseminating agricultural knowledge and promoting rural development. The following are some of the initiative of extension activities through facebook:

Providing timely and accurate weather information is crucial for farmers. KVK Bengaluru Rural shares daily weather updates and forecasts, helping farmers make informed decisions about planting, irrigation, and harvesting. Providing daily weather updates and forecasting information is a crucial service that KVK, can offer to the farming community.

Importance for Farmers

1. Weather significantly impacts planting, irrigation, harvesting, and other farming operations. Accurate forecasts help farmers make informed decisions and minimize risks.
2. Warnings about adverse weather conditions, such as heavy rainfall, storms, or heatwaves, allow farmers to take protective measures.
3. Knowing the expected rainfall helps farmers optimize irrigation, conserving water resources.
4. Timely weather information can help farmers prevent crop damage and livestock losses.

• Credibility of Information

KVK, Bengaluru Rural often obtain weather data from reliable sources, such as: India Meteorological Department (IMD), Agricultural universities and specialized weather forecasting services thereby maintaining the credibility of the information. The information provided typically includes daily and weekly weather forecasts, rainfall predictions, temperature forecasts, warnings about severe weather events and agro-meteorological advisories tailored to specific crops. By sharing daily weather updates and forecasts on their Facebook channel, KVK, Bengaluru Rural, can reach a large number of farmers quickly and efficiently, and, Provide localized weather information relevant to the region and enhance farmers' ability to make informed decisions.

• Daily Agricultural Market Price Updates

Providing daily agricultural market price updates through Facebook is a highly beneficial service that ICAR-KVK, Bengaluru Rural, can offer to farmers. Here's why it's important and how it contributes to their well-being. Farmers need to know the current market prices of their produce to make informed decisions about when and where to sell. This helps them maximize their profits. Providing market price updates increases transparency in the agricultural market, reducing the risk of farmers being exploited by intermediaries. Knowing the prevailing market prices empowers farmers to negotiate better prices for their produce. Access to timely market information can help farmers avoid selling their produce at excessively low prices.

• Video Documentation of Technologies Released by UAS, Bengaluru

ICAR-KVK, Bengaluru Rural, uses Facebook channel to spread the knowledge of latest farm technologies that are developed by UAS Bengaluru. This is done by creating and posting video documentations to the facebook platform. This allows the KVK to reach a wider audience of farmers. Further if provides easy access to information, facilitate interaction and answer farmers' questions. The videos often include, Demonstrations of new farm techniques, Explanations of new crop varieties, Information on innovative agricultural equipment and Expert opinions from researchers, UAS, Bengaluru Videos make it easier for farmers to understand and adopt new technologies.

Accessibility: Farmers can access information from the comfort of their homes any where

Engagement**:** Facebook allows for comments and questions, fostering a two-way communication channel.

• Video Documentation of Farmers' Opinions on Frontline Demonstrations (FLDs)

The video documentation of farmers' opinions on Frontline Demonstrations (FLDs) is a very valuable practice employed by KVK. Farmers' opinions are essential for evaluating the effectiveness of new technologies. Farmer's feedback provides insights into the practical challenges and benefits of adopting these technologies in field conditions and this feedback helps researchers and extension workers refine and improve technologies.

When farmers see and hear positive feedback from their peers, they are more likely to adopt new technologies. Farmer feedback helps KVKs tailor their extension activities to the specific needs of the community. Feedback from FLDs provides valuable data for researchers to improve agricultural technologies. It gives a clear view of the real world results of the work being done by the KVK's, these stories are also shared on Facebook to inspire and educate.

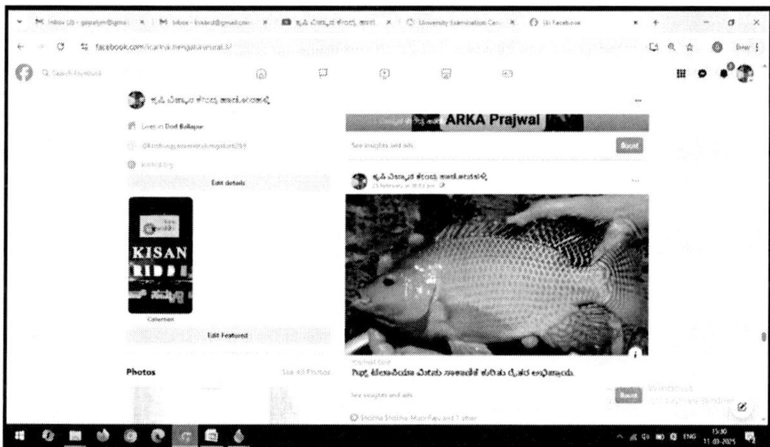

Fig 2. Sharing Farmer's feedback on FLDs through Facebook channel

• Success Stories of Progressive Farmers

Highlighting the achievements of progressive farmers inspires and motivates others, these videos showcase innovative practices, improved yields, and increased income, demonstrating the tangible benefits of adopting new technologies. The video documentation of success stories of progressive farmers is another key activity undertaken by KVK.

Purpose and Importance

- **Inspiration and Motivation:** The videos showcase the achievements of farmers who have successfully adopted new technologies, innovative practices, or diversified their farming systems. They serve as a powerful source of inspiration and motivation for fellow.
- **Demonstrating Real-World Impact:** Success stories provide tangible evidence of the benefits of adopting new agricultural techniques. They show how these practices translate into increased yields, improved income, and enhanced livelihood.
- **Knowledge Sharing:** These videos share the specific strategies and techniques employed by successful farmers allowing others to learn from their experiences.
- **Building Trust and Credibility:** By highlighting local farmers, KVK, builds trust within the community and reinforces its role as a reliable source of information.

• Technology Video Capsules

Short, concise video capsules are created to explain specific technologies or practices in a clear and engaging manner. These capsules focus on key aspects, making it easy for farmers to understand and implement new techniques. Technology video capsules on Facebook, produced by KVK, are a concise and effective way to disseminate agricultural information.

Benefits for Farmers

Sharing video capsules through Facebook has its own advantages which includes

1. Farmers can learn about new technologies at their own pace and from the comfort of their homes.
2. The visual demonstrations make it easier for farmers to understand and apply new techniques.
3. Facebook allows for the rapid dissemination of information ensuring that farmers have access to the latest technologies.

Facebook allows KVK to reach a wider audience of farmers, Video capsules are more engaging than text-based posts, leading to increased interaction with farmers as they provide a cost-effective way to disseminate agricultural information.

- **Regular Activities of KVK**

Videos documenting workshops, training programs, field days, and other activities of KVK are uploaded to YouTube, keeping farmers informed about ongoing activities and opportunities.

II. YouTube

The ICAR KVKs, are increasingly leveraging YouTube as a powerful tool for agricultural extension.

Technology Dissemination: KVKs create and upload videos demonstrating new agricultural technologies, practices and techniques developed by institutions like UAS, Bengaluru. These videos often include practical demonstrations, expert explanations and visual aids to enhance understanding.

Farmer Education and Training: YouTube serves as a platform for delivering educational content to farmers, covering topics such as crop management, pest control, livestock rearing, and soil health. They publish recordings of training sessions, workshops, and expert lectures, making the KVK to accessible a wider audience.

Showcasing Success Stories: KVKs document and share success stories of progressive farmers who have adopted innovative practices and achieved positive results. These videos inspire and motivate other farmers to adopt similar techniques.

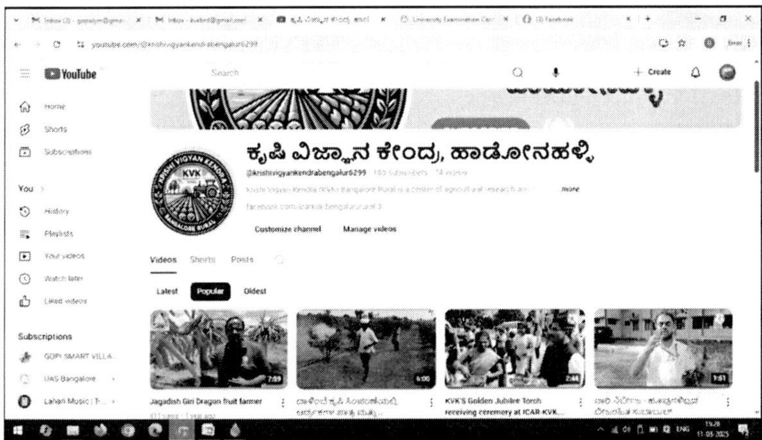

Fig 3. Homepage of YouTube channel of ICAR-KVK, Bengaluru Rural

Documenting Frontline Demonstrations (FLDs): YouTube is used to share videos of FLDs, capturing farmers' experiences and opinions on new technologies. This provides valuable feedback and builds trust among the farming community to adopt the feasible technologies.

Sharing KVK Activities: KVKs upload videos showcasing their regular activities such as field days, exhibitions, and community outreach programs. This helps to raise awareness of the KVK's services and engage with the local community.

Benefits of Using YouTube

The advantages of using YouTube are :

1. Videos provide a visual and engaging way to learn, making it easier for farmers to understand complex concepts.
2. YouTube allows farmers to access information at their own pace and convenience.
3. The platform enables KVKs to reach a large audience, including farmers in remote areas.
4. YouTube facilitates the sharing of knowledge and best practices among farmers.

III. Instagram

Instagram's focus on visual content makes it ideal for showcasing agricultural practices and success stories. The ICAR-KVK Bengaluru rural use Instagram to share UAS, Bengaluru technology videos, leveraging the platform's visual strength. Short reels and posts showcase practical demonstrations, highlighting new farming techniques and crop varieties. This visual approach simplifies complex information making it accessible to farmers. "Behind-the-scenes" glimpses of research and farmer testimonials build trust. By using Instagram, KVKs reach a wider audience, especially younger farmers, and encourage the adoption of innovative agricultural practices. Concise, engaging videos ensure quick comprehension, promoting technology dissemination and improving agricultural productivity.

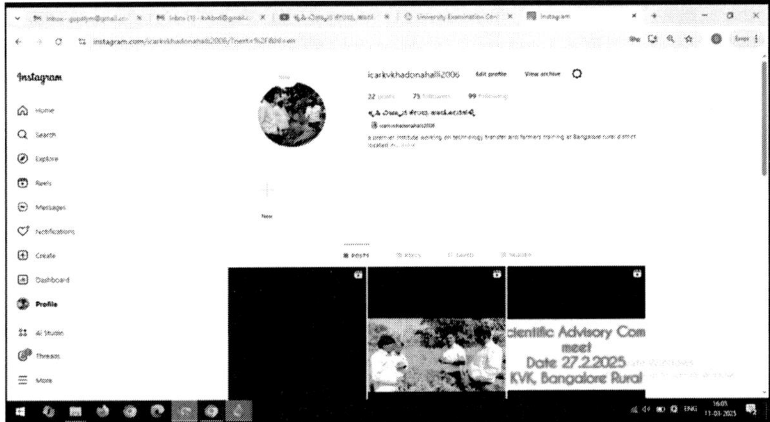

Fig 4. Homepage of Instagram account of ICAR-KVK, Bengaluru Rural

Additionally, KVK, Bengaluru Rural, utilizes Instagram to share farmer feedback on Frontline Demonstrations (FLDs). Short video clips capture authentic testimonials, showing first-hand experiences with new technologies. These visuals demonstrate practical benefits, fostering trust and encouraging adoption. Farmers' voices validate research, highlighting real-world results. Instagram's format allows for quick, impactful sharing, reaching a broad audience. This strategy enhances extension efforts by providing relatable evidence of technology effectiveness, directly from farmers themselves. This approach strengthens community engagement and promotes the adoption of improved agricultural practices. Further, the KVK also uses Instagram to showcase success stories of progressive farmers, leveraging the platform's visual power to inspire. Short, engaging videos and images highlight innovative practices and tangible results. These stories demonstrate the real-world impact of adopting new technologies, motivating other farmers. By visually portraying successful agricultural endeavours, KVK fosters a sense of community and encourages knowledge sharing. Instagram's reach allows these stories to inspire a wider audience, promoting the adoption of best practices and driving agricultural progress. These visuals serve as powerful testimonials, proving the effectiveness of modern farming techniques.

IV. X-formerly Twitter

X's brevity and real-time updates make it suitable for quick dissemination of information. KVK Bengaluru Rural uses X for sharing videos of UAS, Bengaluru's released technologies on X for quick dissemination. Concise video clips and links to detailed content provide farmers with rapid access to vital agricultural advancements.

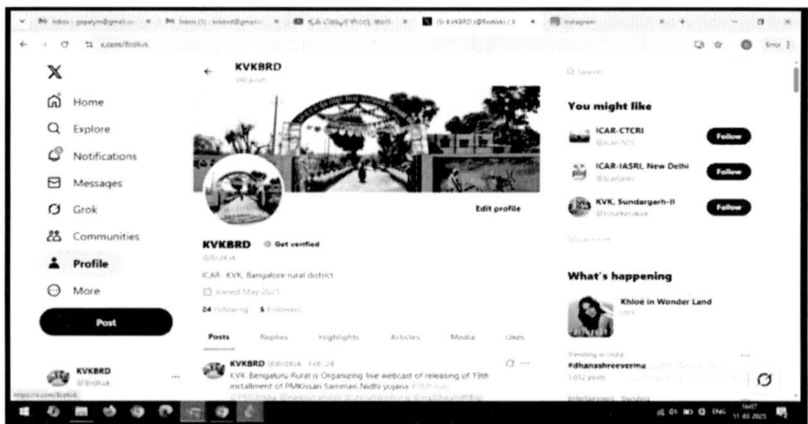

Fig 5. Homepage of X (Twitter) account of ICAR-KVK, Bengaluru Rural

This strategy leverages X's real-time nature to deliver timely updates on new technologies, seed varieties, and equipment. short, impactful visuals convey key information, encouraging farmers to explore further. By utilizing X, KVK ensures widespread awareness of cutting-edge agricultural solutions, promoting adoption and enhancing productivity in the region. Further, KVK uses X to share concise videos of farmers' opinions on Frontline Demonstrations (FLDs). These short clips offer quick, impactful testimonials, showcasing real-world experiences with new agricultural techniques. Sharing these videos via X allows for rapid dissemination of farmer feedback, highlighting the practical benefits of FLDs. This approach provides timely validation of technologies, encouraging wider adoption.

By leveraging X's immediacy, KVK ensures farmers receive prompt and relatable insights, fostering trust and promoting effective agricultural practices within the community. Addition to this, shares success stories of progressive farmers on X through short, impactful videos. These clips highlight innovative practices and tangible results, offering quick inspiration and practical insights. By leveraging X's immediacy, KVK rapidly disseminates these stories, fostering a sense of community and promoting knowledge sharing. These concise videos act as powerful testimonials, demonstrating the effectiveness of modern agricultural techniques. Sharing these success stories encourages wider adoption of best practices, contributing to improved livelihoods and agricultural progress within the region. KVK uses X and to share updates on its regular activities, often linking to Instagram content. This allows for quick, concise announcements of workshops, training, and field days. X serves as a rapid notification system, directing followers to detailed visual documentation on Instagram. By bridging these platforms, KVK ensures timely dissemination of information about its events and services. X's brevity complements Instagram's visual richness, providing a comprehensive overview of KVK's engagement. This cross-platform strategy maximizes reach, keeping the community informed and encouraging participation in KVK initiatives.

Conclusion

ICAR-KVK Bengaluru Rural's strategic use of social media platforms has significantly enhanced its extension efforts. By leveraging the unique strengths of each platform, the KVK effectively disseminates agricultural technologies, shares valuable information, and builds a strong community of farmers. This multi-platform approach ensures that farmers have access to timely and relevant information, empowering them to adopt innovative practices and improve their livelihoods. The KVK's dedication to digital extension serves as a model for other agricultural institutions seeking to bridge the digital divide and promote agricultural development.

References*

1. Abbott, J. 2017. Introduction: Assessing the social and political impact of the internet and new social media in Asia. *J. Contemp. Asia* 43, 579-590. doi:10.1080/00472336.2013.785698.
2. Chukwuere, J.E. and Chukwuere, P.C. 2017. The impact of social media on social lifestyle: A case study of university female students. *Gender Behav,* 15, 9966-9981.
3. Iwamoto, D. and Chun, H. 2020. The emotional impact of social media in higher education. *Int. J. High. Educ.* 9, 239-247. doi: 10.5430/ijhe. v9n2 p239.

* The content of this chapter was prepared in consultation with above references.

15

Entrepreneurship Development Initiatives of Institute of Baking Technology and Value Addition, UAS, Bengaluru

***Savita S. Manganavar*[1], *Y.N. Shivalingaiah*[2], *A. Ashwini*[3] *and Prakruthi N. Raj Gangadkar*[4]**

[1]*Coordinator & Head, IBT&VA,* [2]*Director of Extension,* [3,4]*Assistant Professors, IBT&VA, Directorate of Extension, UAS, Bengaluru*

Email: deuasBengaluru@gmail.com

Abstract

The Institute of Baking Technology and Value Addition, University of Agricultural Sciences, Bengaluru has made significant contributions to bakery and food processing. The institute offers comprehensive training in bakery technology and value-added food processing over the years, it has developed over 20,000 entrepreneurs, primarily from Karnataka but also from states like Bihar, Uttar Pradesh, Kerala, Andhra Pradesh, Telangana, Assam, Maharashtra and New Delhi. In 2008, the Institute received the National Productivity Council Award from the Ministry of Food Processing Industries, Government of India, acknowledging its excellence in food processing training. The institute has established bakery production units at Parappana Agrahara Central Prison and the Women Central Jail in Tumakuru with CSR funding from BHEL, Bengaluru. It has also conducted training programs for the wives of military personnel at the MEG Center in Halsur, Bengaluru promoting economic empowerment. The Further, institute has facilitated the establishment of 16 women-run bakeries in rural areas and provided training to tribal populations enhancing their social integration and economic stability. Through initiatives like the "Learn Bake Earn" concept, the institute has empowered 145 stakeholders to pursue self-entrepreneurship. It has developed over 250 entrepreneurs, created employment for more than 80 individuals and offers technical support to over 1,000 entrepreneurs annually. The annual income of entrepreneurs trained by the institute ranges from Rs. 25,000 to Rs. 22,00,000, collectively generating an estimated revenue of around Rs. 4406.4 lakh each year.

Keywords: *Bakery, Entrepreneurs, Food Processing, Income generation and Value addition.*

Introduction

Bakery industry is one of the largest segments in the food processing sector in India. The bakery industry offers huge opportunities for growth, innovation, and job generation. This industry is separated into three categories such as bread, biscuits, cakes and pastries. The global bakery industry reached a market value of Rs. 42.07 lakh crore in 2024 and around 1.2 lakh crore from the Indian market as India is the second largest producer of biscuits after the USA. Thus, India is a key player internationally and with the entrepreneurial spirit of Indian companies and individuals. Thus, bakery sector is one of the most sustainable business opportunities.

Changing consumer habits, increased urbanization, improved standards of living, convenience needs of dual income families and lifestyle are shaping the bakery industry in India. Part of a global trend, there is greater demand for healthier products and alternatives, particularly when it comes to bakery products which are now more commonly consumed daily as compared to being considered as a treat in the olden days. With high consumption rates, consumers are demanding healthy options in bakery products seeking for gluten-free products or goods made with alternative ingredients such as multigrain or whole-wheat flour. Other than healthier options, most of the youths and children are demanding for new flavours, Fibre dense products and rich experiences, making diversified innovations. The industry is consolidating this demand by strengthening bakery products with the aim of satisfying a rapidly growing appetite and creating awareness among consumers with health concepts and nutritional value.

While there is a demand and appetite for bakery products in both urban and rural areas, and thus, recently the bakery segment has increasingly matured to a great extent. The industry is generally divided into organised and unorganised, with more than 2,000 organised or semi-organised bakeries and more than 1,000,000 unorganised bakeries. However, the Indian bakery industry faces certain challenges. Operational efficiency is a major issue in the industry, as is the lack of technology and skilled workers.

Bakery products due to high nutrient value and affordability, are an item of huge consumption. Due to the rapid population rise, the rising foreign influence, the emergence of a female working population and the fluctuating eating habits of people, they have gained popularity among people, contributing significantly to the growth trajectory of the bakery industry.

Bakery holds an important place in food processing industry and is a traditional activity, with respect to Millet bakery products, consumers are demanding newer options and the industry has been experiencing fortification of bakery products in order to satiate the burgeoning appetite of the health-conscious people. The bakery industry has achieved third position in generating revenue among the processed food sector. The growth rate of bakery and other value added products has been tremendous in the both urban and rural areas. The sector has indicated promising growth prospects and has been making rapid progress.

Although there are obstacles which are causing losses, there has been a boom in entrepreneurial endeavors in the bakery industry in India. Home baking has always been a popular pursuit, but with new technological innovations individuals have been able to monetize their efforts. At every level in the bakery industry in India there are challenges and opportunities. While the rise of local home bakers is being encouraged to promote entrepreneurship in bakery sector. The skills are developed through long term and short term capacity building programs in Healthy baking and Value addition sector.

The Institute of Baking Technology and Value Addition is a prestigious institute started during 1968 working under the Directorate of Extension, University of Agricultural Sciences, Bengaluru. It was started during 1968 as Baking school sponsored by US Wheat Associates, New Delhi, with the main objective to impart skill oriented training programs in bakery.

The Institute of Baking Technology and Value Addition has been imparting training in bakery and food processing for potential entrepreneurs, professional bakers, industrialists, SHG members, youths, housewives and other interested groups, with an aim to create and upgrade technical competency, to develop confidence regarding different stages of food processing and product development, to develop entrepreneurship and to create self-employment opportunities. Presently the institute is equipped with semi-automatic plant to conduct training programs on large scale manufacturing of bakery products and infrastructure for conducting research activities.

Since 2006, the unit also includes "Food processing" which is a major component of training for processing and preparation of value added products using different agricultural produce such as fruits, vegetables, cereals, millets, milk, soybean, spices etc. The main aim of these training programs is to promote technical competency and self-entrepreneurship encouraging secondary and tertiary agriculture for doubling the income.

Institute of Baking Technology and Value Addition is working with a Vision to provide trusted and valued knowledge, consistently through training programs to obtain high product quality, create employment opportunity, provide outstanding service to all customers (bakery entrepreneurs, housewives & bakery owners) to be Indian best managed bakery and processing consultant institute. It has a Mission to apply strong and consistent production practices and principles in pursuit of our service and valued commitments; to generate economically sustainable value added Bakery owners and Bakery suppliers.

The Institute is Functioning with the following Objectives Significant Achievements of Institution

During 2008, the Institute of Baking Technology and Value Addition bagged "National Productivity Council award" from Ministry of Food Processing Industries, Government of India, New Delhi for excellent functioning on food processing. The Institute of Baking Technology and Value Addition has the credentials of developing more than 20,000 entrepreneurs both in bakery and value addition industry since its establishment, and has contributed for the nutritional and economic empowerment of the general public in general and rural masses in specific. It has created awareness about Baking Science, Value addition and Entrepreneurship development among more than two lakh individuals for past 20 years. The participants were major from Karnataka state and also from other states such as Bihar, Uttar Pradesh, Kerala, Andhra Pradesh, Telangana, etc.

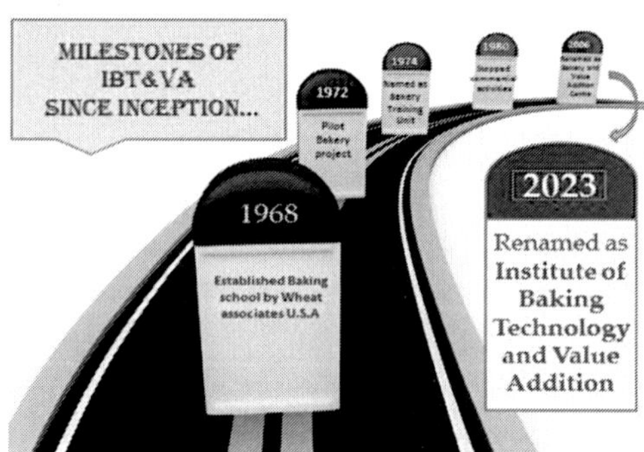

Fig 1: Milestones of IBT & VA since Inception

The Technologies Addressed

The institute has been imparting training in bakery since last five decades for potential entrepreneurs, professional bakers, industrialists, SHG's members, youths, housewives and other interested groups, etc. to develop entrepreneurship and to create self-employment opportunities. Presently, the institute is equipped with pilot scale plant with experts to impart training of bakery products on large scale manufacture of bakery products by step by step process. Annually around 1000 trainees are trained at the Institute of Baking Technology and Value Addition on product development, standardization on bulk production, packaging, marketing, etc. The institute is also engaged in development of novel products, byproduct utilization of farm produces for effective utilization and additional income, and designing the curriculum for need-based courses.

Table 1. Special products developed at Institute of Baking Technology and Value Addition

Rich chocolate cake	Fancy bread
Masala Doughnut	Whole-wheat bread
Sweet doughnut	Soup sticks
Chocolate doughnut	Easter bread / bun
Aloo bun	Garlic bread / loaf
Garlic bread	Croissants
Multigrain bread	Mexican roll
Millet bread	Focaccia
Black forest cake	Poppy seed knots
Pineapple pastry	Danish pastry
Fresh fruit cake	Ohma biscuit
Mini pizza	Brown bread
Vegetable roll	Bread rolls
Egg roll	Sweet bun/ fruit bun
Stuff puffs	Swiss Roll
Egg puffs	Rich Plum Cake
Dilpasad	Red velvet cake
Bombay kara	Eggless cake
Burger	Millet jaggery cake
Banana cake	Millet dosa mix
Carrot cake	Millet baked nippatu
Danish pastry	Moringa chutney powder
Masala bun	Palak bites
Mexican roll	Ragi ambli mix
All types of Decorated cakes	Millet magic drink
Queen Cake	Millet Idli mix
Puff Pastry	Amla candy
Prolines Biscuit	Amla squash
Chocolate Crunch	Amla murabha
Orange Custard Biscuit	Amla chithrana spice mix
Custard Salt Biscuit	Millet ladoo

Cherry Knobs	Ragi ladoo
Almond Cookies	Flaxseed ladoo
Nut Rings	Jackfruit products

Table 2. Millet based Bakery products developed at Institute of Baking Technology and Value Addition

Little Millet Bread	Multi millet sesame cookies
Little Millet Tea Bun	Little millet pumpkin cookies
White ragi based sweet and salt biscuit	Millet based cake premixes
Millet based crackers	White ragi rusk with stevia
Millet based Masala crackers	White ragi based coconut cookies with stevia
Millet jaggery cake	Millet based eggless cake

Table 3. On-going research work at Institute of Baking Technology and Value Addition

Kasuri methi crackers	Whole-wheat bread
Baked samosa	Multigrain bread
Baked Kachori	Baked momos
Sour dough bread	Millet bread
Bambolins	chocolate mousse Pastry
Bagels	Molten chocolate cake
Rolled cakes	Flourless chocolate
Rawa cakes	Filled doughnuts
Truffle cakes	Tresleches
Mascarpone	chocolate mousse Pastry (Eggless)

Table 4. Products available for Commercialization developed at Institute of Baking Technology and Value Addition

Name of the product	
Enriched Ragi Vermicelli	Coconut flour Sweet and salt Biscuit
Ragi Vermicelli with madhunashinil leaves powders	Coconut Flour Salt Biscuit
Ragi Vermicelli with amruthaballi leaves powder	Coconut Flour Nutri strips
Ragi Vermicelli with germinated fenugreek seeds powder	Coconut Flour Ladoo
Ragi Vermicelli with Jamun seeds powder	Enriched Gluten free cookies
Ragi Vermicelli with Ashwaganda root powder	Gluten free Little millet biscuit
Coconut flour Cookies	Gluten free Foxtail millet biscuit
Coconut Flour Masala Biscuit	Low fat Little millet biscuit with WPC
Coconut Flour Cake	Low fat Foxtail millet biscuit with WPC
Coconut Flour Rusk	

Extension Management Practices

Initiatives of IBT & VA

1. Providing fundamental knowledge on bakery and value addition
2. Imparting knowledge on bakery raw materials
3. Hands on training in preparation of bakery and value added products
4. Imparting knowledge on food hygiene, sanitation and safety
5. Costing and economics of bakery and value-added products
6. Promotion of bakery and value addition industry for economic empowerment
7. Training on handling of semi-automatic plant
8. Development and technical support to potential entrepreneurs
9. Supply of quality human resource to the food industries
10. Ruralisation of bakery and value addition industry

Other Services of IBT & VA

- Off campus training programs
- Orientation/Refresh programs
- Awareness programs
- Sponsored training programs
- Collaborative training programs
- Exhibitions/ display in Fair/Melas
- Internship for students
- Consultancies/Technical support
- Research in Bakery and Value addition
- Incubation services for startups

Opportunities and Challenges Documented for Bakery/Food Processing Industry

Opportunities

- Innovation and Product Diversification: Bakeries that focus on innovation, new product development, and expanding their product portfolios to cater to evolving consumer preferences will have a competitive edge and drive market growth.
- Increasing Awareness of Healthier Options: The growing health-consciousness among consumers presents opportunities for bakeries to introduce and market healthier bakery products, including whole grain options, gluten-free alternatives, and natural ingredients.

Challenges

- Heavy Competition (focus on innovation, product differentiation, and maintaining consistent quality)
- Ingredient Sourcing: Ensuring a consistent supply of high-quality ingredients can be a challenge
- Changing Consumer Preferences: As consumer preferences continue to evolve, bakeries must stay attuned to the demand for healthier options, gluten-free products, and vegan alternatives.
- Premium and Artisanal Offerings: There is a growing demand for premium and artisanal bakery products including specialty cakes, pastries, and artisan bread. Consumers are willing to pay a premium for unique and high-quality baked goods.
- Online Bakery: The rise of e-commerce platforms and online food delivery services has created opportunities for bakeries to reach a wider customer base. Online bakery platforms, offering convenience and a diverse range of products, are gaining popularity.
- Customization and Personalization: Customers increasingly seek customized bakery products tailored to their preferences and dietary needs. Bakeries that offer customization options and personalized experiences have a competitive advantage.

Other Challenges

- Fluctuation in raw material prices
- Price-sensitivity
- Application of new technology
- Sales promotion techniques
- Distribution strategies
- Government legislations
- Present scenario: Impact of post – COVID
- Mode of buying
- Consumption pattern
- Lack of awareness on support through govt schemes
- Lack of knowledge on compliance requirements

Output

Activities of the Institute during last Five Years

Table 6. List of various activities carried out during last five years

Activities	Number of Training programs	Number of participants
4 Weeks Baking course	46	876
14 weeks certificate course	12	318
Sponsored trainings	41	985
Short courses	94	2167
Orientation programs	54	3046
Collaborative training	34	1068
Internship/Implant training	20	225
Off campus training	55	3352
Special events	36	12500
Exhibitions	26	10400
Technical consultancy	600	600
Advisory services	1800	1800
Total	2818	37337

The Institute of Baking Technology and Value Addition had organized 12 programs of 14 Weeks Certificate Course in Bakery Technology (commercial production) for 318 participants and 46 programs of 4 weeks Bakery courses (Domestic courses & small scale production) for 876 participants. Besides, the Institute has conducted 94 on-campus short courses of different durations for 2167 participants, 54 Orientation Programmes for 3046 members, 34 Collaborative/sponsored trainings for 1068 members, 20 Internship/Implant trainings for 225 members, 55 Off-campus trainings for 3352 participants and 41 sponsored training programs were organized under projects for 985 participants. Overall, 12,037 participants were exposed to various trainings.

The institute has organized 36 programs on special days such as World Food day, World Milk Day, Mahila Kisan diwas etc., for 12,500 members, 26 important events such as krishimela, cake exhibition and Novel bakery products - exhibition cum sale for more than 10,400 participants. The institute has provided more than 600 technical consultancies and 1,800 no. of advisory services, to create awareness regarding institute objectives in promoting food processing and entrepreneurship by providing professional production and management skills so as to promote availability of technical personnel in the field of bakery and food processing industry. More than 25,000 participants visited these exhibitions and special events during last five years and procured useful information.

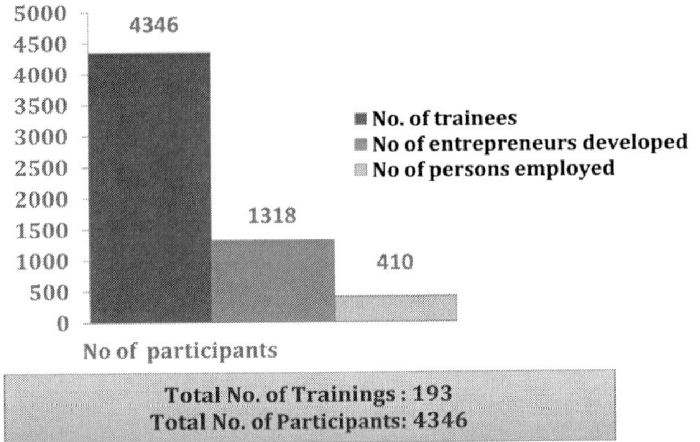

Fig 2. Impact of training programs at the institute during last five years

Impact of Training programs

Among various training programs conducted in the Institute, 193 programs were intensive focused trainings such as Fourteen Weeks Certificate Course in Bakery Technology (Commercial production), 4 weeks Bakery courses (Domestic courses & small scale production), short courses on value addition and sponsored trainings were conducted for 4346 participants and more than 400 participants have procured jobs in various food processing units & more than 1300 have started their own ventures as emerging entrepreneurs over the past five years.

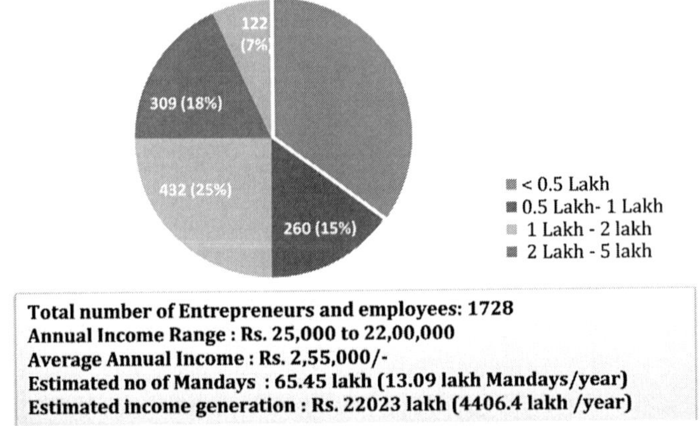

Fig 3. Annual income of entrepreneurs & employees developed by the institute during last five years

Annual Income of Entrepreneurs and Employees

The institute has supported 1728 participants with technical guidance and providing consultancies to become sustainable Entrepreneurs and Employees of Food Processing Industries with annual income ranging between Rs 25000 to Rs 22, 00,000 and average annual income of Rs 2, 55,000 with estimated no of mandays of 13.09 lakh days per year and overall to an extent of 65.45 lakh mandays for past five years.

Documentation and Outcome/ Impact over the Past 20 Years

- This Institute has the credentials of developing more than 20,000 entrepreneurs both in bakery and value addition industry since its establishment
- Created awareness about Baking Science, Value addition and Entrepreneurship development among more than 2 lakh individuals from past 20 years.
- The participants enrolled were, major from Karnataka state and also from other states such as Bihar, Uttar Pradesh, Kerala, Andhra Pradesh, Telangana, Assam, Maharashtra, New Delhi, etc.
- Contributed for the Nutritional and economic empowerment of the common public in general and rural masses in specific.
- During 2008 the Bakery Training Unit bagged "National Productivity Council Award" from Ministry of Food Processing Industries, Government of India, New Delhi for excellent functioning on food processing.
- Established Bakery Production Unit at Parappana Agrahara, Central Prison, Bengaluru with an income of Rs 20,000 per day
- Established Bakery production unit at Women Central jail, Tumkur with the CSR fund of BHEL, Bengaluru
- Conducted Training programs for Wives of Jawans of MEG center, Halsur, Bengaluru for encouraging economic empowerment
- Established 16 (sixteen) exclusively women run Bakeries in rural areas
- Bakery Training programs to the Tribals to make them lead their life securely in the society.
- 145 Stakeholders were promoted for self-entrepreneurship through "Learn Bake Earn" concept
- More than 250 number of Entrepreneurs are being developed every year

- Employment opportunities created for more than 80 individuals every year
- Technical and continued support for more than 1,000 entrepreneurs every year
- Annual Income ranged from Rs. 25,000 to Rs 22,00, 000 per entrepreneur
- Estimated total revenue generation is around Rs. 4,406.4 lakh per year
- Estimated number of man days created is around Rs. 13.09 lakh days per year

Conclusion

IBT & VA is promoting in development of entrepreneurs with economic sustainability for a comfortable living. Further, it will also provide impetus to the food processing industry to add value in terms of economic and social livelihood. The indirect impact of it will improve the self-esteem and self-confidence which further adds for improving their attitude towards life and health.

Implications/ Strategic Recommendations

- The bakery industry in India is poised for substantial growth in the coming years, driven by changing consumer preferences, rising disposable incomes, and growing urbanization, changing lifestyles and technological advancements.
- Despite challenges such as competition, ingredient sourcing, and infrastructure requirements, the industry offers significant opportunities for established players and aspiring entrepreneurs.
- By leveraging innovation, catering to evolving consumer demands, and maintaining product quality, companies can unlock the tremendous potential that the Indian bakery market holds.

References

1. https://www.researchandmarkets.com/report/india-bakeries
2. https://www.imarcgroup.com/indian-bakery-market
3. https://www.google.com/search?q=bakery+industry+market+size+in+india&sca
4. https://www.google.com/search?q=wheat+production+in+india&sca

16

Remodeling of Training Module at Extension Education Institute, Hyderabad: Integrating Scientific Techniques, Methodologies and Digital Tools

***M. Jagan Mohan Reddy*[1]*, M. Srinivasulu*[2]*, Ch. Venu Gopala Reddy*[3]*, M. Preethi*[4]*, V. Ravinder Naik*[5]*, K. Aruna*[6] *and M. Yakadri*[7]**

[1]*Director and University Head (Agril. Extn.), EEI (Southern Region),* [2]*Coordinator, Electronic Wing,* [3]*Dean of Student Affairs,* [4,5]*Professor, EEI (Southern Region),* [6]*Assistant Director of Extension,* [7]*Director of Extension, PJTAU, Rajendranagar*

Email: depjtau@gmail.com

Abstract

The Extension Education Institute (EEI), Hyderabad, has adopted a scientific approach to remodeling its training module by integrating innovative techniques, models, and digital tools to enhance the effectiveness of training programs. The initiative involves the structured use of online tools across 13 stages, including need assessment, registration, pre- and post-evaluation, and feedback collection at three intervals. The training follows David Kolb's Experiential Learning Cycle and the Kirkpatrick Model of Training Evaluation, ensuring continuous improvement and impact assessment. The training programs, designed for middle-level officers of Southern states and union territories, incorporate 60-70% practical content, field visits, and interactive learning methodologies such as quizzes, simulations, role plays, and management games. This type of innovative training model enhances the effectiveness of the training programs.

Keywords: *Digital tools, Experiential learning, Extension education, Kirkpatrick model, Participatory learning and Training module*

Introduction

In the ever-evolving landscape of agricultural extension education, the need for dynamic and interactive training approaches has become paramount.

Traditional training methods often focus heavily on theoretical content with minimal engagement. To bridge this gap, EEI Hyderabad has adopted a scientific and structured approach to training delivery. This initiative focuses on digital integration, experiential learning, and robust evaluation techniques to enhance knowledge retention and practical application among participants.

About the Technologies Addressed

The initiative leverages various digital tools and scientific models to enhance training effectiveness. These include:

- Online tools (13 stages): Used for participant registration, need assessment, evaluation, and feedback collection.
- David Kolb's Experiential Learning Cycle: Ensuring hands-on learning through experience, reflection, conceptualization, and application.
- Kirkpatrick Model of Training Evaluation: A four-level approach to measuring training effectiveness through reaction, learning, behavior, and results.
- Simulation Exercises and Management Games: Enhancing decision-making skills and real-world problem-solving.
- Quizzes and Role Plays: Encouraging active participation and knowledge reinforcement.
- Fishbowl Technique and Focused Group Discussion: Facilitating peer-to-peer learning and critical thinking.

Extension Management Practices

a) Initiatives

- **Remodeling training modules by incorporating scientific techniques and models:** To enhance the effectiveness of training at EEI, Hyderabad, remodeling of training modules by integrating scientific techniques and models is essential. This involves adopting evidence-based instructional methods such as the Kirkpatrick Model for training evaluation, Bloom's Taxonomy for structured learning, and experiential learning models for hands-on engagement. Data-driven decision-making, statistical analytics, and behavior change models can further refine training content, ensuring practical application in real-world scenarios. Additionally, integrating participatory approaches like simulations, case studies, and interactive assessments fosters deeper understanding. By continuously updating modules with scientific advancements, EEI, Hyderabad equips extension professionals with cutting-edge knowledge and practical skills for impactful field applications.

- **Digitalization of training processes using Online toolsat various stages:** The Online toolsused at different stages of the training program are enclosed with the document in point no.11.

Adoption of Experiential Learning and Kirkpatrick Evaluation Models

Application of Kolb's Experiential Learning Cycle in Designing Training Modules at Extension Education Institute, Hyderabad

Training programs at the Extension Education Institute (EEI) play a crucial role in equipping professionals, rural youth, and other stakeholders in the agricultural and allied sectors with necessary skills and knowledge. To enhance the effectiveness of these programs, it is essential to incorporate evidence-based learning theories. One such framework is Kolb's Experiential Learning Cycle, which provides a structured approach to adult learning by emphasizing the importance of experience, reflection, conceptualization, and experimentation. By integrating Kolb's model into training module design, EEI ensures a more engaging and impactful learning experience.

Kolb's Experiential Learning Cycle in Training Design

Kolb's model consists of four stages: Concrete Experience, Reflective Observation, Abstract Conceptualization, and Active Experimentation. Each stage plays a crucial role in fostering a comprehensive learning experience. When designing training programs at EEI, modules are structured to align with these four stages to cater to different learning styles effectively.

Concrete Experience: Engaging Learners Through Real-world Scenarios

- Training sessions begin with direct experiences relevant to the participants' professional backgrounds.
- Activities such as field visits, hands-on demonstrations, role-playing exercises, and case studies are employed to immerse learners in real-life agricultural and extension challenges.
- Example: A session on 'Sustainable Farming Practices' has a with a visit to an organic farm, allowing trainees to observe and participate in sustainable agricultural techniques firsthand.

Reflective Observation: Encouraging Critical Thinking

- After engaging in direct experiences, learners are given time to reflect on their observations. This stage enables them to connect their experience with prior knowledge and identify gaps in understanding.

- Techniques such as group discussions, reflective journaling, and peer sharing are employed.
- Example: Following the organic farm visit, participants engage in a guided discussion where they share their insights on the benefits and challenges of sustainable farming, analyzing their observations from different perspectives.

Abstract Conceptualization: Developing Theoretical Understanding

- At this stage, learners conceptualize their experiences and develop theoretical insights.
- Trainers introduce scientific principles, research findings, and expert knowledge to help participants build a solid conceptual foundation.
- Example: A lecture on soil health management is integrated after the farm visit and reflection session, linking real-world observations with theoretical principles like soil fertility enhancement and microbial activity.

Active Experimentation: Applying Learning to Real-World Contexts

- The final stage involves applying newly acquired knowledge and skills to real-life situations, allowing learners to test their ideas and refine their understanding through practice.
- Activities such as simulation exercises, project-based learning, and implementation of action plans in trainees' respective fields are incorporated.
- Example: Participants could design and implement small-scale experiments in their own communities, such as setting up a composting unit based on the principles learned during the training. These learning outcomes are presented by participants through their back home planning session.

Aligning Training with Kolb's Learning Styles

Kolb's model also identifies four learning styles: Diverging, Assimilating, Converging, and Accommodating. To maximize engagement and effectiveness, training modules at EEI incorporate elements that appeal to each of these styles:

- Diverging Learners (Feeling and Watching): Include group discussions, storytelling, and creative problem-solving activities.
- Assimilating Learners (Watching and Thinking): Use research-based lectures, case studies, and structured analysis of agricultural issues.

- Converging Learners (Doing and Thinking): Integrate hands-on workshops, data-driven problem-solving exercises, and technical skill-building sessions.
- Accommodating Learners (Doing and Feeling): Provide opportunities for field visits, experiential projects, and decision-making tasks.

Practical Implementation at EEI

- Module Design: Each training program is structured to cycle through the four stages of experiential learning, ensuring holistic skill development.
- Facilitation Techniques: Trainers employ a mix of instructional methods, including hands-on practice, theoretical discussions, and real world applications.
- Assessment Strategies: Evaluation done at EEI includes reflective assessments, practical demonstrations, and project-based learning outcomes rather than just traditional written tests.
- Continuous Improvement: Feedback from participants is incorporated to refine future training programs, ensuring alignment with diverse learning preferences.

By integrating Kolb's Experiential Learning Cycle and Learning Styles into training design, the Extension Education Institute enhances the effectiveness of its programs. This approach ensures that participants not only gain theoretical knowledge but also develop critical thinking and practical problem-solving skills that are essential for their roles in agricultural extension and rural development. A well-structured training module that leverages experiential learning principles will lead to more engaged learners and improved knowledge retention, ultimately contributing to the advancement of sustainable agricultural practices and rural development initiatives.

Increased focus on participatory and practical learning methodologies

Enhancing Training Effectiveness through Participatory Approaches at EEI, Hyderabad

The Extension Education Institute (EEI), Hyderabad, has been at the forefront of capacity building and knowledge dissemination for officers in the agriculture and allied sectors. To ensure the effectiveness of training programs, EEI incorporates participatory learning approaches such as the Fishbowl Technique, Focused Group Discussions (FGDs), Role Plays, Presentations, Panel Discussions, and Management Games. These methods not only engage participants actively but also facilitate experiential learning, critical thinking, and skill development, making the training more impactful.

Fishbowl Technique: Encouraging Open Dialogue

The Fishbowl Technique is a highly interactive method used at EEI, Hyderabad to encourage inclusive discussions on complex issues. In this technique, a small group of participants sit in an inner circle discussing a topic, while the remaining participants form an outer circle, observing and listening. Observers can take turns entering the discussion circle, ensuring diverse perspectives are shared. This method is particularly useful in deliberating policy issues, extension challenges, and case studies related to agricultural development. It promotes deep listening, active participation, and thoughtful exchange of ideas among trainees.

Focused Group Discussions (FGDs): Gaining Insights from Practitioners

FGDs at EEI, Hyderabad serve as an effective tool for exploring perceptions, experiences, and attitudes of participants on specific agricultural and extension-related topics. Facilitators guide discussions with open-ended questions, encouraging participants to share real-life experiences and challenges faced in the field. This method provides valuable qualitative insights into agricultural extension policies, adoption of technologies, and capacity-building needs, which help in designing targeted interventions. FGDs also foster peer learning, as participants gain knowledge from their counterparts' experiences.

Role Plays: Enhancing Practical Application

Role-playing exercises are incorporated into EEI's training modules to simulate real-world extension scenarios. Participants enact different roles such as agricultural officers, progressive farmers, policymakers, or stakeholders in a rural development scenario. By stepping into different roles, participants develop empathy, communication skills, and problem-solving abilities. This technique is especially beneficial in training programs focused on conflict resolution, leadership, negotiation, and farmer advisory services. It allows trainees to practice theoretical concepts in a controlled environment before applying them in the field.

Presentations: Strengthening Communication and Knowledge Sharing

EEI encourages participants to prepare and deliver presentations on key topics related to agri and allied areas, extension services, and policy frameworks. Presentations help participants articulate their thoughts clearly, structure their ideas logically, and enhance public speaking skills. This method also allows knowledge exchange, as trainees share best practices and innovative solutions from their respective regions. Feedback from facilitators and peers further refine their presentations and analytical skills, preparing them for professional engagements.

Panel Discussions: Integrating Expert Insights

Panel discussions at EEI bring together subject matter experts, policymakers, and experienced practitioners to deliberate on pressing issues in agri and allied areas and rural development. Moderated discussions provide a platform for trainees to engage with experts, ask questions, and gain deeper insights into complex topics. This approach fosters a multidisciplinary perspective, helping participants understand the interconnectedness of various factors influencing agricultural extension and policy implementation. Panel discussions also stimulate critical thinking and expose participants to different viewpoints thus equipping them with a broader knowledge base.

Management Games: Building Leadership and Decision-Making Skills

Management games are an innovative training approach used at EEI to enhance leadership, teamwork, and strategic decision-making abilities among participants. These games simulate real-world agricultural extension challenges, requiring participants to collaborate, analyze scenarios, and make informed decisions within a time frame. Activities such as resource allocation exercises, crisis management simulations, and team-building challenges help participants develop problem-solving skills in a dynamic and engaging manner. Management games also promote creativity, adaptability, and resilience which are key attributes for extension professionals working in diverse field conditions.

Impact of Participatory Approaches on Training Outcomes

Incorporating participatory learning methods in EEI's training programs has yielded significant benefits. These approaches:

- Enhance engagement and retention: Interactive techniques ensure that participants are actively involved, leading to better comprehension and long-term retention of knowledge.
- Promote experiential learning: By simulating real-life situations, trainees gain practical experience that prepares them for field challenges.
- Encourage collaborative learning: Group activities foster peer-to-peer knowledge exchange and create a network of professionals who can support each other beyond the training program.
- Improve soft skills: Communication, leadership, negotiation, and teamwork skills are honed through role plays, discussions, and presentations.

- Facilitate real-time problem-solving: Participants analyze case studies and scenarios, enabling them to apply theoretical knowledge to real-world situations.

EEI, Hyderabad, continues to refine its training methodologies by integrating participatory approaches that enhance the learning experience. The combination of fishbowl discussions, FGDs, role plays, presentations, panel discussions, and management games ensures that training is not only informative but also engaging and practical. These techniques empower extension professionals with the necessary skills and confidence to drive agricultural innovation and rural development effectively. By prioritizing interactive learning, EEI, Hyderabad reinforces its commitment to building a competent and dynamic workforce for the agriculture and allied sectors.

Training Course Evaluation by Using Kirkpatrick Model

Application of the Kirkpatrick Model in Evaluating EEI Training Programmes

The Extension Education Institute (EEI), Hyderabad, is dedicated to providing high-quality training programs for professionals in agriculture and allied sectors. Evaluating these training programs is crucial to ensuring their effectiveness and alignment with institutional and stakeholder goals. One of the most recognized models for training evaluation is the Kirkpatrick Model, which provides a structured, four-level approach to assess the impact of training programs.

Level 1: Reaction

The first level of the Kirkpatrick Model focuses on the participants' reaction to the training program. At EEI, it is essential to gauge whether trainees find the training relevant, engaging, and useful for their professional roles. Feedback is collected through surveys, interviews, and direct observations at the end of each training session.

To implement this effectively:

- Uses both online and paper-based feedback forms
- Ask participants to rate aspects such as course content, trainers' expertise, training methodologies, and material relevance.
- Include open-ended questions to capture qualitative feedback.
- Encourage honest responses by assuring trainees that their feedback will be used for improvement rather than evaluation.
- Conduct informal verbal feedback sessions during training breaks to capture real-time reactions.

By assessing reaction levels, EEI, Hyderabad makes immediate modifications to the training design, such as adjusting content delivery methods, incorporating more interactive sessions, or refining training materials.

Level 2: Learning

This level evaluates the extent to which participants have acquired knowledge, skills, attitudes, confidence, and commitment from the training program. EEI assesses learning outcomes through pre-and post-training assessments, knowledge tests, and skill demonstrations.

Methods for evaluating learning at EEI include :

- Administering pre-training tests to gauge baseline knowledge.
- Conducting post-training evaluations to measure knowledge gain.
- Using case studies, role-plays, and practical exercises to assess skill development.
- Collecting feedback from trainers on participant engagement and comprehension.
- Encouraging self-reflection through individual learning logs or journals.

By systematically measuring learning, EEI determines the effectiveness of its training content and pedagogical strategies. If gaps are identified, training modules are revised to ensure better knowledge retention and skill enhancement.

Level 3: Behavior

The third level examines whether participants apply the knowledge and skills gained during training in their workplace. For EEI, this means assessing how trainees implement new agricultural extension strategies, improved communication techniques, and updated management practices in their respective institutions or communities.

Key strategies for measuring behavior change include:

- Conducting follow-up surveys and interviews with participants 1 to 3 months after training.
- Engaging supervisors or colleagues to provide feedback on observed changes in participants' work performance.
- Organizing field visits or on-site observations to assess the practical application of training concepts.
- Encouraging participants to submit case studies or success stories on how they have implemented their learning in real-world settings on EEI Blog.

It is essential to recognize that behavior change is influenced by organizational culture, resources, and management support. If barriers to implementation exist, EEI collaborates with stakeholders by organizing workshops and online meetings to address these challenges and facilitate a more enabling environment for trainees.

Level 4: Results

The final level of the Kirkpatrick Model evaluates the overall impact of training programs on institutional and sectoral outcomes. EEI aims to contribute to the professional development of extension officers and the advancement of agricultural and rural development initiatives. Measuring results at this level involves linking training outcomes to Key Performance Indicators (KPIs).

Ways to assess results include:

- Tracking improvements in agricultural productivity, extension outreach, or stakeholder engagement as a result of training interventions, through follow up studies.
- Analyzing policy changes or institutional improvements influenced by trained professionals.
- Measuring increased efficiency in agricultural advisory services provided by trainees.
- Assessing the economic and social impact of extension programs facilitated by EEI trainee alumni.

While measuring long-term results is challenging, EEI, Hyderabad uses a combination of quantitative metrics (such as increased farm yields or reduced post-harvest losses) and qualitative insights (such as improved farmer satisfaction and knowledge dissemination) to evaluate success.

The Kirkpatrick Model provides a comprehensive framework for evaluating EEI's training programs by assessing reaction, learning, behavior, and results. By systematically applying this model, EEI Hyderabad ensures that its training programs remain impactful, relevant, and aligned with the needs of agricultural and allied sector professionals. Furthermore, continuous evaluation helps EEI refine its curriculum, enhance trainer effectiveness, and contribute meaningfully to capacity-building efforts in the region.

b) Reach

Middle-level Extension officers from 6 South Indian States including Odisha and Union Territories (Puducherry, Lakshadweep, Andaman and Nicobar Islands):

With ever growing technological interventions in agricultural sector and rapidly growing challenges with globalization and environmental changes, constant upgradation of skills and professional competencies of middle level extension officers is an everyday requirement. Extension officers are the knowledge bearers for the farmers and there is a need for them to stay relevant – therefore, upscaling themselves on the latest technologies in agriculture and allied fields is the only way out.

The success of middle level functionaries lies in successfully and effectively transferring knowledge to the farming community. Well planned, systemically organized, need based trainings to enhance the knowledge, skills and attitude of middle level functionaries go a long way in improving the productivity and effective discharge of their professional duties.

With the sole aim of catering to various training needs for middle level functionaries, Extension Education Institutes were initiated in four parts of India. The EEI at Rajendra nagar, Hyderabad was established in 1962. Under the aegis of Ministry of Agriculture & Farmers Welfare, Govt of India and works under the administrative control of Professor Jayashankar Telangana Agricultural University. It provides training needs for six southern states namely Andhra Pradesh, Karnataka, Kerala, Tamil Nadu, Telangana, Odisha and three Union territories namely Andaman and Nicobar, Lakshadweep and Puducherry islands.

On-campus and off-campus mode training courses

Based on the training needs, the programs are categorized as on campus and off campus trainings. For on campus trainings, the EEI has seminar halls equipped with hi-tech audio-visual capabilities, boarding and lodging, dining hall, sports and well-equipped library facilities. Off campus trainings are conducted in the respective departmental training institutes of the respective locations.

The major mandate of EEI is to impart trainings related to building knowledge and skills on extension and training methodologies, best possible training techniques, conduct action-research and provide consultancy to development agencies. The training needs of middle level functionaries of all the states & union territories are identified during the regional workshops and training need assessment meetings. Thus, a training calendar is prepared and distributed to all the stakeholders.

Continuous engagement with trainees through post-training evaluations

c) Output

- **Enhanced Participation and Engagement in Training Sessions:** The incorporation of participatory learning approaches such as group discussions, role plays, case studies, and management games significantly enhances engagement during training sessions at EEI,

Hyderabad. These interactive methods encourage active involvement from participants, fostering a dynamic learning environment where they can share experiences and perspectives. By utilizing digital tools, multimedia content, and gamification strategies, training sessions become more immersive and enjoyable. Engaging methodologies also cater to different learning styles, ensuring better knowledge retention. When participants feel involved in the learning process, they are more motivated to apply new concepts and contribute meaningfully to discussions, ultimately improving their overall learning experience.

- **Increased Application of Learned Skills in real-world Scenarios:** The success of any training program is measured by how effectively participants implement the acquired knowledge in practical settings. By incorporating hands-on activities such as field visits, case studies, and experiential learning models, EEI ensures that trainees can bridge the gap between theory and practice. Real-life problem-solving exercises and scenario-based trainings allow participants to develop critical thinking and decision-making skills, making them more confident in applying their learnings in agricultural extension work. Continuous mentoring and post-training follow-ups further support trainees in adapting new techniques, ensuring long-term impact and sustainable improvements in their professional roles.
- **Systematic Assessment and Documentation of Training Impact:** A structured assessment framework is crucial for measuring the effectiveness of training programs. EEI employs systematic evaluation techniques such as pre and post training assessments, participant feedback surveys, and performance tracking to gauge the impact of its training initiatives. The use of the Kirkpatrick Model enables multi-level evaluation, assessing reaction, learning, behavioral changes, and final results. By maintaining comprehensive documentation, EEI can analyze training outcomes, refine methodologies, and enhance future programs. Systematic data collection also helps in reporting training effectiveness to stakeholders, securing further support, and ensuring continuous improvement in the training processes for agricultural extension professionals.

d) Outcome/Impact

Improved Knowledge Retention and Skill Application among Participants

One of the primary goals of training programs at EEI, Hyderabad, is to ensure that participants retain the knowledge gained and effectively apply it in their professional roles. To achieve this, modern training methodologies such as experiential learning, hands-on demonstrations, case studies, and role-playing exercises are integrated into sessions. These techniques reinforce theoretical concepts through practical engagement, enabling learning more memorable and applicable. Additionally, post-training follow-ups, refresher courses, and mentorship programs help participants reinforce their understanding and apply newly acquired skills in real-world scenarios. By incorporating statistical analytics and assessment tools, EEI can track individual learning progress and identify areas requiring further intervention. Enhanced retention and application of knowledge ultimately lead to improved problem-solving capabilities among agricultural extension professionals, resulting in more effective service delivery to farmers and rural communities.

Strengthened Extension Service Delivery Through Well-Trained Officers

A well-trained agricultural extension workforce is crucial for enhancing rural livelihoods and improving farm productivity. EEI's training programs focus on equipping extension officers with the latest scientific advancements, digital tools, and participatory methodologies to engage effectively with farming communities. By providing hands-on experience with innovative extension approaches such as Participatory Rural Appraisal (PRA), Farmer Field Schools (FFS), and ICT-based advisory services, officers become more proficient in delivering timely and relevant agricultural advice. Strengthened extension services translate into better technology transfer, increased adoption of improved farming practices, and enhanced problem-solving capabilities at the grassroots level. Additionally, EEI emphasizes leadership and communication skills, enabling extension officers to build trust and establish effective knowledge-sharing networks among farmers.

Increased Adoption of Digital and Scientific Tools in Agricultural Extension Programs

The integration of digital tools and scientific techniques in agricultural extension is transforming the way information is disseminated to farmers. EEI, Hyderabad, promotes the use of mobile apps, AI-driven decision-support systems, remote sensing, and GIS technologies to enhance agricultural advisory services. Training programs include sessions on using digital platforms for real-time weather forecasting, soil health assessment, precision farming, and market intelligence. By equipping extension professionals with these tools, EEI facilitates a faster and more efficient transfer of knowledge to farming communities. Additionally, the use of online learning platforms and virtual training modules expands access to training resources, allowing a wider audience to benefit from EEI's expertise, especially during covid and post covid times. The increased adoption of digital and scientific tools ensures that extension services remain relevant, efficient, and accessible to farmers in an increasingly technology-driven agricultural landscape.

Evidence-Based Decision-Making in Training Effectiveness and Improvements

A data-driven approach to training evaluation is essential for continuous improvement and alignment with emerging needs in the agricultural sector. EEI utilizes systematic assessment frameworks such as the Kirkpatrick Model to measure training effectiveness across multiple dimensions—reaction, learning, behavior, and results. By collecting and analyzing participant feedback, performance metrics, and field impact assessments, EEI identifies areas for enhancement in training content, methodology, and delivery. This evidence-based approach ensures that future training programs are tailored to address real-world challenges faced by extension professionals and farming communities. Additionally, collaboration with research institutions and industry experts allows EEI to integrate the latest scientific findings into its training modules. By adopting a structured evaluation and feedback mechanism, EEI strengthens its ability to design impactful, relevant, and forward-thinking training programs that drive meaningful improvements in agricultural extension services.

Conclusion

The remodeling of EEI Hyderabad's training module represents a significant shift toward scientific and digital-based extension education. By integrating digital tools, experiential learning techniques, and robust evaluation models, the institute has enhanced the effectiveness of its training programs. The initiative serves as a model for other extension training institutes aiming to improve knowledge transfer and practical skill development.

Implications/Strategic Recommendations

Expansion of Digital Learning Platforms for Wider Accessibility

Digital learning platforms have transformed the way training is conducted, making it more accessible, flexible, and engaging for a wider audience. At the Extension Education Institute (EEI), Hyderabad, expanding digital learning platforms is crucial to reaching a larger group of learners, including agricultural officers, extension personnel, and rural development professionals.

By leveraging online learning management systems (LMS), mobile applications, and virtual classrooms, EEI provides training sessions that accommodate diverse learning preferences and geographic limitations. For instance, integrating webinars, interactive online courses ensure and content sharing platforms EEI ensures that trainees can access materials at their convenience, reducing the constraints of physical attendance.

Additionally, EEI strengthens its digital presence by offering multifarious training programs tailored to different regional needs. This inclusivity allows extension professionals from various sector backgrounds to benefit from specialized trainings. Moreover, digital platforms enable real-time monitoring and evaluation of trainees' progress, allowing trainers to offer personalized guidance and support. With the rise of social media adaptive learning, EEI customizes training experiences based on individual learning patterns. Incorporating discussion forums, peer-to-peer learning networks, and gamified training elements further enhances engagement and retention. Ultimately, the expansion of digital learning platforms ensures that EEI's training programs are accessible to a broader audience, leading to a more significant impact on client departments across client states.

Strengthening Collaboration with Educational and Research Institutions for Updated Training Content

Collaboration with educational and research institutions plays a pivotal role in keeping training programs at EEI relevant, innovative, and evidence-based. By forming partnerships with universities, agricultural research centers, and policy think tanks, EEI ensures that its training content is continuously updated to reflect the latest advancements in agricultural practices, extension methodologies, and rural development strategies.

For instance, partnering with institutions like the MANAGE, NAARM, NIPHM, SAMETIs, ni-msme, PJTAU etc., facilitate the exchange of expertise, research findings, and best practices. These collaborations also help in developing specialized training modules focused on emerging topics such as climate-resilient agriculture, digital extension tools, and sustainable farming practices.

Moreover, joint research initiatives can lead to data-driven insights that enhance training methodologies. Guest lectures, expert panel discussions, and collaborative projects with academia is bringing fresh perspectives to EEI's training programs. Another critical aspect is engaging with international agricultural research bodies to incorporate global best practices through its webinars on contemporary aspects with global partnerships.

Regular collaboration with research institutions ensures that EEI's training remains dynamic and responsive to evolving challenges in the agri and allied sectors. This approach enhances the credibility of the institute and strengthens its role as a key knowledge hub for extension professionals.

Regular Adaptation and Refinement of Training Methodologies Based on Feedback Analysis

Continuous improvement of training methodologies is essential to ensure effectiveness and relevance. At EEI, regular feedback collection from participants, trainers, and stakeholders help refine and adapt training programs to meet changing needs and expectations.

Post-training surveys, one-on-one interviews, and focus group discussions are used to assess the effectiveness of different training modules, participatory approaches, and learning outcomes. A systematic approach to analyzing feedback is being done that enables EEI to identify strengths, gaps, and areas for improvement in its training delivery.

For example, if trainees find a particular session too theoretical, EEI incorporates more hands-on exercises, field demonstrations, or case studies. If feedback suggests that digital content is not engaging enough, interactive elements such as quizzes, real-life scenarios, and virtual simulations are introduced. Additionally, incorporating trainers' feedback ensures that teaching methodologies remain learner-centric and adaptable to different learning styles.

By institutionalizing a culture of continuous improvement, EEI ensures that its training programs remain cutting-edge, practical, and impactful for extension professionals and rural development practitioners.

Incorporation of Emerging Technologies like AI-Driven Analytics and Virtual Reality Simulations

Emerging technologies such as Artificial Intelligence (AI) and Virtual Reality (VR) are revolutionizing training methodologies, making learning more immersive, data-driven, and impactful. EEI integrates these technologies to enhance the quality and efficiency of its training programs.

AI-driven analytics can be used to track participants' progress, assess learning patterns, and provide personalized recommendations. For example, AI-powered chatbots that offer instant support and answer queries from trainees, making learning more interactive are used in the form of quizzes and simulations. AI-based assessments help gauge knowledge retention and skill application, allowing trainers to modify content accordingly.

Virtual Reality (VR) simulations provide experiential learning opportunities for trainees. For instance, VR-based scenarios that simulate real-life agricultural extension challenges, such as pest management, soil health assessment, or disaster preparedness are shown with hands-on approach through mobile apps that allow trainees to practice decision-making in a risk-free environment, leading to better knowledge retention and skill development.

Another application of technology is the use of data visualization tools and predictive analytics to interpret trends in agricultural extension services. By integrating IoT-enabled sensors and real-time data feeds, EEI tries to equip trainees with insights into smart farming techniques, climate adaptation strategies, and digital extension services through its field visits to state of art organizations like INCOIS, CCMB, ICRISAT etc.,

The adoption of emerging technologies not only modernizes EEI's training approach but also prepares extension professionals to effectively leverage digital tools in their fieldwork. This ensures that agricultural extension services remain relevant, efficient, and aligned with the evolving demands of the sector. By focusing on these key areas like expanding digital learning platforms, strengthening collaborations, refining training methodologies, and integrating emerging technologies, EEI Hyderabad enhances the effectiveness and reach of its training programs. These initiatives will contribute to building a more skilled, knowledgeable, and tech-savvy workforce in agricultural extension and rural development.

On boarding iECHO platform in conducting online training programs

The novel iECHO platform was facilitated to deliver the content to participants with less dose and more frequency iECHO is a digital platform that provides easy access to training services and enable the participants to present their cases on the selected subject to get the solutions from the trainers. It operates on hub and spokes model.

References*

1. Agricultural Extension in South Asia (AESA). (2022, May 20). *Webinar on 'Professional competence of rural development advisors' organized by the Society of Extension Professionals & Extension Education Institute, Hyderabad, India*. https://aesanetwork.org/webinar-on-professional-competence-of-rural-development-advisors-organized-by-the-society-of-extension-professionals-extension-education-institute-hyderabad-india-20-may-2022/
2. Extension Education Institute, Hyderabad. (n.d.). *EEI blog*. https://www.eeihyd.org/eeiblog/
3. Kolb, D. A. 1976. *The learning style inventory: Technical manual*. McBer.
4. Kolb, D. A. 1981. Learning styles and disciplinary differences. In A. W. Chickering (Ed.), *The modern American college* (pp. 232–255). Jossey-Bass.
5. Kolb, D. A. 1984. *Experiential learning: Experience as the source of learning and development* (Vol. 1). Prentice-Hall.
6. Kolb, D. A., & Fry, R. 1975. Toward an applied theory of experiential learning. In C. Cooper (Ed.), *Studies of group process* (pp. 33–57). Wiley.
7. Kolb, D. A., Rubin, I. M., & McIntyre, J. M. 1984. *Organizational psychology: Readings on human behavior in organizations*. Prentice-Hall.
8. Kurt, S. 2014. *Kirkpatrick model: Four levels of learning evaluation* (8p).

9. MindTools. 2017. *Kirkpatrick's four-level training evaluation model.* Retrieved from https://www.mindtools.com/pages/article/morale.htm
10. National Institute of Agricultural Extension Management (MANAGE). 2024. *Training calendar 2024-25.* https://www.manage.gov.in/trainings/trgCalendar-2024-25.pdf
11. Professor Jayashankar Telangana State Agricultural University. (n.d.). *Extension Education Institute (EEI).* https://www.pjtsau.edu.in/eei.html

* The content of this chapter was prepared in consultation with above references.

Annexure-I

Application of David Kolb's Experiential Learning Model in Designing the Training Program

The training program on 'Promotion of Farmer Producer Organisations in Agri and Allied Sectors' has been designed using David Kolb's Experiential Learning Model, which consists of four stages: Concrete Experience, Reflective Observation, Abstract Conceptualization, and Active Experimentation. These principles ensure that participants engage in a holistic learning process, enhancing their understanding and application of knowledge.

1. Concrete Experience

The training begins with real-world case studies and field experiences shared by experts from established Farmer Producer Organizations (FPOs). Participants interact with successful FPO leaders to gain firsthand insights into their functioning, challenges, and strategies. This immersive experience provides a strong foundation for learning.

2. Reflective Observation

After exposure to concrete experiences, participants engage in structured discussions, group reflections, and interactive Q&A sessions. They analyze their observations, compare them with their prior knowledge, and identify key takeaways. This stage helps in internalizing lessons from real-world examples and encourages critical thinking.

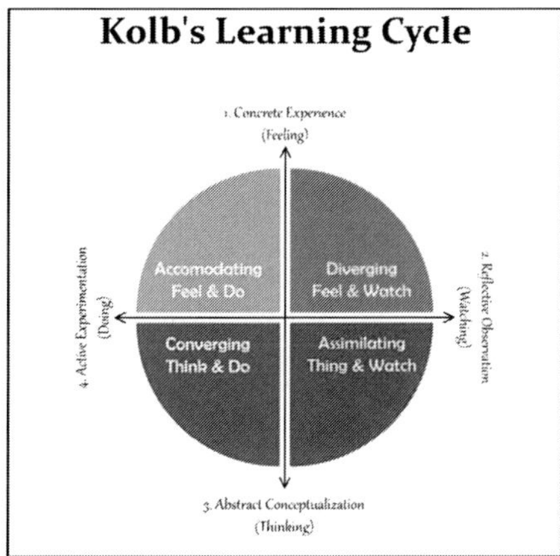

3. Abstract Conceptualization

Participants are then introduced to theoretical concepts, government policies, regulatory frameworks, and best practices related to FPOs. Experts from institutions like MANAGE and EEI present models and frameworks that help in structuring the knowledge gained from previous stages. This stage bridges practical experiences with academic understanding, facilitating deep learning.

4. Active Experimentation

To reinforce learning, participants engage in hands-on activities such as preparing business plans for FPOs, strategizing marketing approaches, and conducting SWOT analyses. Role-playing exercises, scenario planning, and simulation games further enable them to apply theoretical knowledge in practical decision-making. These activities prepare them to implement their learning in real-life situations effectively.

By integrating Kolb's Experiential Learning Model into the training design, the program ensures that participants not only gain knowledge but also develop critical skills for establishing and managing successful Farmer Producer Organizations in the agri and allied sectors.

The Kirkpatrick Evaluation Model

Effectiveness of the Training Programme on Values and Work Ethics for Organisational Excellence using Kirkpatrick Model

Values and Work Ethics are crucial for organisational sustainability and excellence. Recognizing their importance, the Extension Education Institute (EEI), Hyderabad, conducted an online on-campus training program from 18-22 January 2022. A total of 51 officers from Agriculture, Horticulture, Fisheries, Animal Husbandry, and allied sectors participated. The training effectiveness was evaluated using Kirkpatrick's Model across four levels:

Level 1: Reaction

- 100% of trainees rated sessions as excellent.
- 75% preferred offline training; 60% suggested a shorter 3-day duration.
- 90% cited stress management, ethical decision-making, appreciative inquiry, positive attitude, and values & ethics as key learnings.
- Overall course ratings: Excellent (54.9%), Very Good (23.5%), Good (21.6%).

Level 2: Learning

- A pre- and post-training assessment showed a 78% increase in knowledge.

Level 3: Behavior

A post-training survey (11 respondents) revealed

- 75% applied values and ethics in professional and personal life.
- 73% practiced ethical decision-making in the workplace.
- 100% found stress management techniques useful, with varying success levels.
- 91% implemented appreciative inquiry and positive attitude techniques.
- 55% adapted to changes effectively, and 45% practiced time management successfully.
- 64% adhered to moral values post-training.

Level 4: Results

Follow-up calls with respondents indicated

- Stress and time management techniques improved efficiency.
- Anger management helped in emotional control.
- Some trainees became resource personnel at the district level.
- Improved self-appraisal scores and recognition from superiors.

The Kirkpatrick Model gives organizations a framework/reference to bucket data together to identify the impact a program or initiative has on an overall organization.

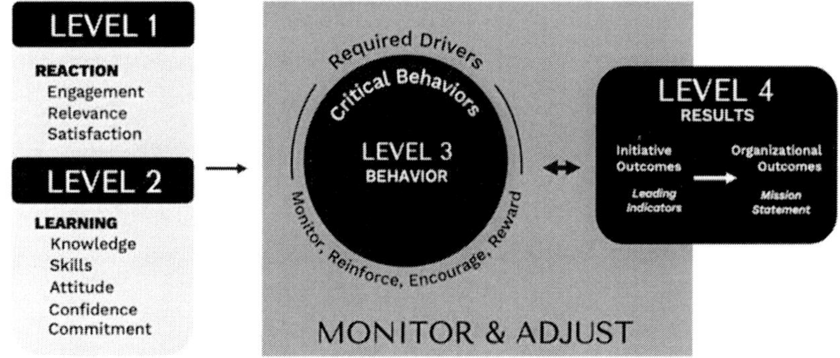

Conclusion

The training significantly enhanced participants' understanding and application of values and work ethics, positively impacting their professional and personal lives. The structured evaluation confirmed the training's effectiveness in fostering ethical behavior, stress management, and improved work efficiency.

Online data collection forms

1. Proforma for the Course Coordinators (after 1 Month)

Proforma for the Course Coordinators ; To evaluate individual training programmes (On/Off/Online)- after one month

Email *

Name of training programme *

Nature of the training programme On/ Off/ Online/ Collaborative

Client Dept. /s attended

2. Proforma for the Trainee participants (after 1 Month) 01.08.2024

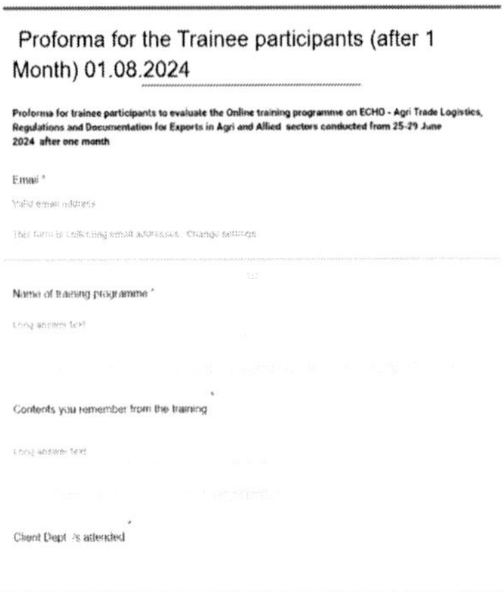

Proforma for trainee participants to evaluate the Online training programme on ECHO - Agri Trade Logistics, Regulations and Documentation for Exports in Agri and Allied sectors conducted from 25-29 June 2024 after one month

Email *

Name of training programme *

Contents you remember from the training

Client Dept. /s attended

3. Proforma for the Course Coordinators (after 3 Months)

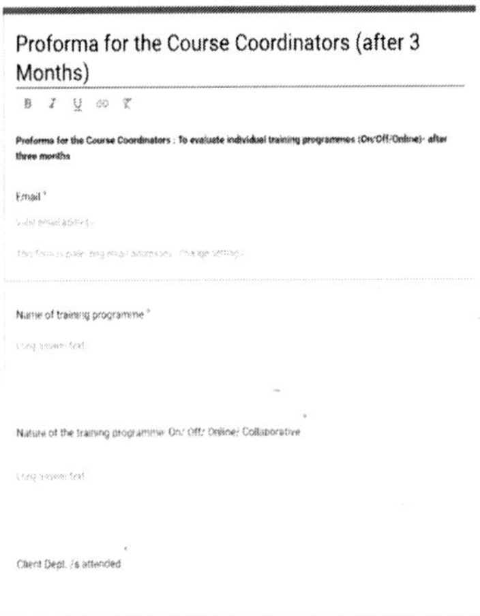

Proforma for the Course Coordinators ; To evaluate individual training programmes (On/Off/Online)- after three months

Email *

Name of training programme *

Nature of the training programme On/ Off/ Online/ Collaborative

Client Dept. /s attended

4. Group Level Planning

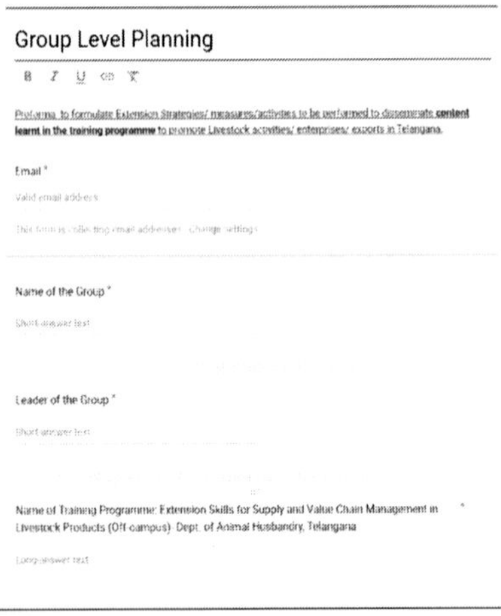

Proforma to formulate Extension Strategies/ measures/activities to be performed to disseminate content learnt in the training programme to promote Livestock activities/ enterprises/ exports in Telangana.

Email *

Name of the Group *

Leader of the Group *

Name of Training Programme: Extension Skills for Supply and Value Chain Management in Livestock Products (Off campus) Dept. of Animal Husbandry, Telangana *

5. 'Proforma for submission of information on Programmes implemented by the faculty of EEI, Hyderabad'.

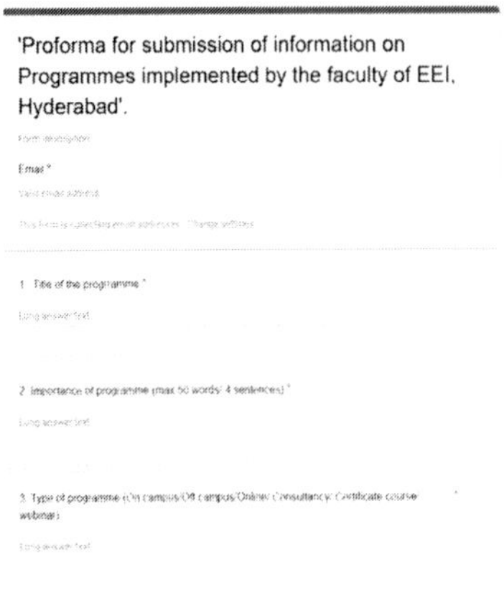

Email *

1. Title of the programme *

2. Importance of programme (max 50 words/ 4 sentences) *

3. Type of programme (On campus/Off campus/Online/ Consultancy/ Certificate course/ webinar) *

6. After 1 Week evaluation form

Proforma for the Course Coordinators ; To evaluate individual training programmes (On/Off/Online)- after one week

Email *

Name of trainee

Designation f trainee

Name of training programme *

7.	After 1 Month evaluation form Form description Email * Valid email address This form is collecting email addresses. Change settings Name of the trainee * Short answer text Designation of trainee * Short answer text Name of training programme * Long answer text Nature of the training programme On/ Off/ Online/ Collaborative *	8.	After 3 Months evaluation form Form description Email * Valid email address This form is collecting email addresses. Change settings Name of the trainee Short answer text Designation of trainee Short answer text Name of training programme * Long answer text Nature of the training programme On/ Off/ Online/ Collaborative *
9.	Effectiveness of Skill Training Program Form description Name * Short answer text Age * Short answer text Gender * Male Female	10.	Case Study as a Training Method Case Study Research Design Questions Email * Valid email This form is collecting emails. Change settings 1. Which of the following studies is used for clinical analysis * A. Correlational Study B. Survey Study C. Case Study D. Trend Study 2. Which of the following is similar to Case Studies * A. Longitudinal Studies B. Social Surveys
11.	Action Research- Nutri Smart Villages (Dr.M.Preethi) General information Village * Short answer text Date of interview * Month, day, year Name of the Respondent * Short answer text Contact number * Short answer text	12.	Feedback on Field visit - On campus Trg. Prog. on "Soft Skills for personal and professional excellence" from 30th January to 3rd February, 2024 Form description Name of the Participant * Short answer text email id * Short answer text Mobile No. * Short answer text

13.

Before Training Evaluation - On Campus training "Digital Marketing Strategies for Agri and Allied Sectors" from 20th to 24th February, 2024

Before Training Evaluation - On Campus training "Digital Marketing Strategies for Agri and Allied Sectors" from 20th to 24th February, 2024

1. Name of the Applicant *

Short answer text

2. Designation *

Short answer text

3. Email address *

Short answer text

4. Educational Qualification *

14.

Post Evaluation - Off Campus -" Leadership & Personality Development Skills for Professional Excellence" From 18th to 21st February, 2025

Post Evaluation - Off Campus -" Leadership & Personality Development Skills for Professional Excellence"

Email *

Valid email

This form is collecting emails. Change settings

Name *

Short answer text

Designation *

Short answer text

1. Incompatibility of goals leads to

A. Negotiation

15.

Registration Form - Off Campus Training Program on "Training of Trainers(ToT)-Direct Training Skills and Teaching methodology " From 10th to 13th March, 2025

Registration Form - Off Campus Training Program on "Training of Trainers(ToT)-Direct Training Skills and Teaching methodology " From 10th to 13th March, 2025

Name of the Applicant *

Short answer text

Designation *

Short answer text

Mobile No *

Short answer text

16.

Pre Evaluation - Off Campus -" Training of Trainers(ToT)-Direct Training Skills and Teaching methodology " From 10th to 13th March, 2025

Form description

Email *

Valid email

This form is collecting emails. Change settings

Name *

Short answer text

Designation *

Short answer text

1. What is the primary objective of management games in a training setting?

A) To entertain participants

Organizing the on-campus training program

Organizing the Off campus training program

Conducting Webinar

Conducting Ice breaking exercise to understand the participants each other

Role play performance by trainees during the training program

Organizing consultancy proramme on "Gender Sensitization"

Organizing Collaborative training program

Conducting Certificate course for the Terrace Gardening aspirants

Skill Development Training program of Garden Keeper- Exhibition by participants

Follow-up studies to assess the impact of the EEI training program in Telangana State

17

The Trailblazer in Frontline Extension to Reach the Unreached

M. Jagan Mohan Reddy[1], M. Srinivasulu[2], Ch. Venu Gopala Reddy[3], V. Ravinder Naik[4], M. Preethi[5], K. Aruna[6] and M. Yakadri[7]

[1]Director and University Head (Agril. Extn.), EEI (Southern Region), [2]Coordinator, Electronic Wing, [3]Dean of Student Affairs, [4,5]Professor, EEI (Southern Region), [6]Assistant Director of Extension, [7]Director of Extension, PJTAU, Rajendranagar

Email: depjtau@gmail.com

Abstract

India's magnificent journey from the state of being food deficit to food surplus is praise worthy. The role of agricultural research organizations and farm Universities in achieving this daunting task cannot be under estimated. Farm universities work with the major mandates of teaching, research and extension. Agriculture is a state subject and it is a known fact that over the years, the ratio of extension officers to farmers is very wide not only in India but across various other parts of the world owing to varied reasons. We are living in the era where, there is no dearth of information, but the problem is getting the quality information which is relevant and to the level of meeting the demands of individual farmers.

The Professor Jayashankar Telangana Agricultural University (PJTAU) in pursuit of addressing the issue of reaching the unreached, has made institutional innovations in the extension landscape as part of its front-line extension efforts. The "Rythu Nestham" Live Video conference facility by the University to offer real time solutions through digital platform was initiated on March 6th, 2024 connecting the Department of Agriculture managed Rythu Vedikas across the state involving multiple stakeholders from Horticulture, Veterinary universities and State department of Agriculture and allied sectors. The University has introduced an innovative student radio program "Chenukaburlu" on 26th January, 2015 to upgrade the students' knowledge on agriculture and allied sectors, enhancing their confidence and communication skills and to inculcate a sense of commitment towards service to the farmers. The radio program, also aims to keep the farmers abreast on the latest developments in farming practices and technologies in *vernacular language. To provide right information at right time in the hands of the farmers, the university has initiated a dedicated YouTube Channel on 30th August, 2017 for showcasing latest agricultural technologies to the farmers as PJTAU Agricultural Videos, with 93,300 Subscribers, 85 lakh viewers so far, developed about 608 videos on major crops is gaining popularity amongst the farmers and is a role model to many other SAUs. To supplement the extension efforts of the KVK and to reach the unreached, one of the KVKs of the University has initiated an Innovative Farmers Network (IFN) during 2016 to identify the innovative farmers in the social system, imparting knowledge on maintaining network, to improve capacity of the innovative farmers, to monitor the information sharing mechanism by the innovative farmers (ways and means followed) and to understand the extent of dissemination of technologies by the innovative farmers.*

Keywords: *Agricultural videos, Chenukaburlu, Front line extension, Innovative farmers network and Rythunestham.*

Introduction

Food is the prime necessity of the life without which the existence of the human being may be perhaps dubious. India is agriculture predominant country where majority of its population depends on agriculture directly or indirectly. Agriculture plays a vital role in India's economy. About 54.60 per cent of the total workforce is engaged in agriculture and allied sector activities (Census, 2011). Agriculture and Allied sector accounts for about 18.4 per cent of India's GVA at current prices during 2022-23. Given the importance of the agriculture sector, Government of India has taken several steps for its development in a sustainable manner (Agriwelfare.gov.in/en/Dept).

India with massive network of about 63 State Agricultural Universities (SAUs), three Central Agricultural Universities (CAUs), four deemed universities and four central universities with agricultural faculty along with Indian Council of Agricultural Research accords due priority to the agricultural education, research and extension (ICAR Annual report, 2023-24). The role of these universities & ICAR in addressing the farm related problems cannot be overlooked.

As per the All India report on Agriculture Census, 2015-16, India is a country with majority of the farmers being small (17.60 %) and marginal (68.50 %). Agriculture is very diversified with varied agro-climatic conditions which require different practices and strategies to perform. The role of agricultural research and extension in achieving food self-sufficiency is immense. Agriculture is a state subject and various players make agriculture profitable and farming a viable option. But the data indicates that, there is scarcity of extension officers at various levels in India. As per the report of the Committee on Doubling Farmers' Income, 2017 (Volume XI "Empowering the Farmers through Extension and Knowledge Dissemination") reported that, the ratio of Extension Service Provider (ESP) to operational farm holdings is around 1: 1156. Coupled with this, the system has mainly focused on the input supply chain and crop cultivation activities.

In the era of 21st century, innovation is the buzz word, and pervading in all spheres of life. It is astonishing to know that, todays discoveries are outdated tomorrow. In today's rapidly changing world, innovation is crucial for organizations and societies to adapt, grow, and maintain a competitive edge. It drives economic development, addresses complex challenges, and improves the quality of life.

Over the years, agriculture sector has also seen rapid changes in all aspects, and India's efforts in this development pursuit are quite inspiring. Starting from food deficit nation to the nation with surplus food production, the role of agricultural research and extension viz a viz., innovation cannot be overlooked. As a result of bifurcation of Andhra Pradesh in the year 2014, the Professor Jayashankar Telangana Agricultural University was established on 03-09-2014 by an act of legislature. Professor Jayashankar Telangana State Agricultural University (PJTSAU) named in honour and memory of Professor Jayashankar, an eminent educationist and an ardent Telangana ideologue is the only Farm University of Telangana State.

PJTAU has eleven constituent colleges with eight of those devoted to faculty of Agriculture, two to faculty of Agricultural Engineering and Technology and one to faculty of Community Science. In addition, there are twelve constituent polytechnics (eleven in Agriculture and Agricultural Engineering). The research component in the agriculture is addressed by Sixteen Agricultural Research Stations, including three Regional Agricultural Research Stations. As part of front-line extension support to reach the unreached, the University has established about nine District Agricultural Advisory and Transfer of Technology Centres (DAATTCs), eight Krishi Vigyan Kendras (KVKs), one each of Extension Education Institute (EEI), Agricultural Information and Communication Centre (AI&CC), Agricultural Technology Information Centre (ATIC) and Electronic Media Wing spread across the State to cater to various needs of the farming community.

To address the farm problems and make agriculture profitable, the Agricultural University not only strives in research but also a pioneer in the institutional Innovations of University Extension system to cater to the emerging needs of the farming community by means of various innovations such as., Flag method of Extension, Interactive Information Dissemination System, Innovative farmers' networking, Developing Farmer – Master trainers, Interactive information Kiosks, Interactive DVDs, PJTAU YouTube Channel, Yuva Rythu Saagubadi, Chenu Kaburlu – Student radio program, WhatsApp groups. All these have yielded positive benefits to the farming community by addressing the farm related problems. The university is reorienting its innovative extension platforms to accelerate its reach to the farmers. A few of such innovations are presented here under.

I. Rythunestham – A Innovative Digital Platform to Offer Real Time Solutions

"Ryhtu Nestham" Live Video conference aimed to offer real time solutions through Digital Platform was initiated on March 6th 2024 by A. Revanth Reddy, the Hon'ble Chief Minister of Telangana by the Electronic Wing centre of the PJTAU as part of RKVY Project 2023-25. With the outlay of Rs. 96.00 crores implemented by the Dept. of Agriculture, Telangana for connecting all the 2601 Rythu Vedikas of the state, the rythu nestahm live program is an innovative digital platform to offer advisories and solutions to the farmers navigated under the presence of Extension officials and university experts. The innovative extension model is envisaged with the objectives to serve as a Digital Expert System for Scientists-Extension Functionaries-Farmers live interaction at Rythu Vedika's and field level there by resolving problems in real time, to build the capacities of Extension Functionaries in utilizing Digital Platform to reach the unreached.

Unique Features of the Model

- Quick / Real time reach of Farm Solutions and Technologies to farmers and Extension functionaries for better farm decision and Acts as effective feedback mechanism to Scientists and Dept. of Agriculture.
- Reduction in cost, time and human resource spent on extension services by Dept. of Agriculture and Agricultural University.

Stakeholders Involved

Expertise from PJTAU / SKLTHU / PVNRTVU, and agriculture and allied Departments.

Knowledge/Technologies Disseminated

Based on season /crop /weather conditions topics finalized with experts and line department officials, accordingly farm advisories will be given to farmers at Rythu Vedikas and field level.

Strategy: To address the problem of delay in transfer of agro-advisory to farmers. Live interaction programs are being initiated with live discussions between scientists and farmers of crop affected areas.

Outcome: Since the initiation of the program, a total of about 48 episodes were completed on Agriculture/Horticulture/Veterinary and other subjects.

Reach: The model aims about 15,000 farmers per week through 566 Ryhtu Vedikas across Telangana State with need-based farm advisories in real time. The model stands unique with the completion of 48 episodes and reached about 5.50 Lakhs farmers and officials. The model in the long run is expected to empower farming community in terms of Technical know-how, access to inputs, credit, marketing and value addition pertaining to Agril. and allied enterprises.

Hon'ble Chief Minister of Telangana Revanth Reddy and Agricultural Minister with farmers

Experts interaction with farmers interacting with farmers

Fig 1. Glimpses of Rythunestahm live program

II. Chenu Kaburlu - Student Radio Programme

The Professor Jayashankar Telangana Agricultural University has introduced an innovative student radio program "Chenukaburlu" on 26th Janauary,2015. The program aims at upgrading the students' knowledge on agriculture and allied sectors, enhancing their confidence and communication skills and to inculcate a sense of commitment towards service to the farmers. it is also aiming to keep the farmers abreast on the latest developments in farming practices and technologies in vernacular language.

Objectives

- To provide an opportunity to the student community to exhibit their creativity
- To encourage the students to explore innovative ideas in transfer of technology through radio communication
- To provide the students an understanding of real-life situations of farming community on one hand and furnish them with latest developments in the field of agriculture and allied sectors on the other hand.

- To support the farming community with required information from time ti time
- Broadcast on every Wednesday from 1.30 to 2.00 PM through All India Radio Hyderabad 'A' Station – 738 MW program named as Vyavasaya Vignana Tarangini/ Gruha Vignana Tarangini. First of its kind from any State Agricultural University in the country.

Reach of Initiative: The Unique program has a massive coverage with about 529 episodes so far being broadcasted covering the entire State of Telangana.

Output: Transfer of need-based farm advisories by Students in the local dialect and slang there by easy to understand and adopt the Creative skills in delivering Radio talks and Writing Skills in Radio scripts development will improved among Students.

Impact: Creative skills in delivering Radio talks and Writing Skills in Radio scripts development will improved among Students.

Fig 2. Glimpses of Student radio program - Chenukaburlu

III. Pjtau Agricultural Videos

In the present century, which is a century of miracles, the information technology has revolutionized many spheres of life. In this era of information technology, the availability of information is not an issue any more, but the availability of right information at right time is a crux in profitable agriculture. To address the issue of providing right information at right time, the Professor Jayashankar Telangana Agricultural University has initiated a dedicated YouTube channel on 30th August, 2017 for showcasing latest agricultural technologies to the farmers as PJTAU Agricultural Videos. The Agricultural Youtube channel with 93,300 Subscribers and so far, developed about 608 videos on major crops. The Youtube channel is ranked 2nd among SAUs in the India in terms of highest number of subscribers.

Reach of Innovation

Entire State of Telangana and A.P with about 85.00 Lakh viewers.

Output: Transfer of need -based farm advisories in video format by experts of crops in local language. The platform is the best medium to less educated farmers as it works on principle of seeing is believing.

Impact: More adoption of latest technologies by farmers as the YouTube platform provide Real time solutions to farmers field-based problems. The video capsules by the experts of the university serves as the best source of agricultural information to the farming community in times of need during various farm operations. Practicing the advisories by the farmers serves to reduce the cost of cultivation, adopting the innovative practices and thereby increasing the returns.

Fig 3. PJTAU Agricultural Videos Youtube channel

IV. Innovative Farmers Network (IFN) by the KVK, Palem

To address the issue of bridging the gap between farmers and extension officers, and to supplement the extension efforts of the KVK and to reach the unreached, KVK, Palem initiated the Innovative Farmers Network (IFN) during 2016. The Innovative Farmers Network was conceptualised with the objectives of identifying innovative farmers in the social system, imparting knowledge on maintaining network, to improve capacity of the innovative farmers, to monitor the information sharing mechanism by the innovative farmers (ways and means followed) and to understand the extent of dissemination of technologies by the innovative farmers.

Unique features

The model aims at disseminating the location-specific knowledge with innovative farmers (Farmer to Farmer) and reaching the other farmers as it is difficult for any institutions to the reach to all the farmers. The concept is gaining popularity over the years and the identified innovative farmer is a bridge between the KVK and fellow farmers.

No. of Stakeholders involved: Scientists, Innovative Farmers and Farmers. The innovative farmers network inculcates the knowledge as well as motivates the farmers for adoption of new technologies in agriculture. Innovative farmers network members are maintaining sub-networks at village level. Each farmer maintains sub-network with an average 15 -20 farmers. The total no. of farmers 375 (=15x25) were covered and these are the methodologies followed by IFN farmers for dissemination of information to fellow farmers. Various extension activities such as Group discussions, Grama sabhas, Agricultural exhibition, field visits, RAWEP programs, Meetings, extension talks, Electronic channel, press coverages are facilitated by these innovative farmers.

Impact

The impact of the model was studied in terms of information receiving, sharing, processing and disseminating behaviour. It was reported that, under personal – cosmopolite channels, majority of IFN-farmers approaching university scientists, in case of non-IFN farmers they are approaching officials of state dept. of agriculture to acquire the latest information on agril. and allied activities. While, under *impersonal – cosmopolite* channels, majority of them approaching Viewing farm telecasts, in case of non-IFN farmers they are participate Kisan melas and Agriculture exhibitions to acquire the latest information on agriculture and allied activities.

The results of the Item analysis of information storing of IFN respondents indicated that, majority of them resorting through electronic gadgets, in case of non-IFN farmers they are storing in the form of newspapers to store the latest information on agriculture and allied activities. Item analysis of information processing of IFN respondents indicates that, majority of them processing the information through trialability and compatibility, in case of non-IFN farmers the information is processed with the help of complexity and feasibility. Item analysis of information processing of IFN respondents indicates that, majority of them expressed that consulting the scientists in case of non-IFN farmers they are discussing with progressive farmers for processing the latest information on agriculture and allied activities. Item analysis of information dissemination of IFN respondents indicates that, majority of them using social media, in case of non-IFN farmers through personally for dissemination of the latest information on agriculture and allied activities.

Conclusion

We are presently living in the era where there is no dearth of the information and also living in times, where the ratio of farmer to extension officers is high coupled with this, extension is a dynamic subject which needs reorientation time and again to reach the unreached with efficient services. In this pursuit, the efforts by the university extension with innovative models such as Rythu Nestham, PJTAU Youtube channel, Chenukaburlu-Student radio program and Innovative farmers networking by the Professor Jayashankar Telangana Agricultural university are still evolving to meet the dynamic needs of the diversified farming community.

Implications

Agriculture is a dynamic subject which is influenced by various biotic and abiotic factors. Innovation in reach of farmers is indispensable for profitable agriculture. The rapidly evolving nature of agricultural innovation processes in countries like India with majority of farmers being small and marginal and the very same farmer is now caught in the vortex of more serious challenges. These require agricultural extension agencies to transform the roles that previously supported in the dissemination and adoption of innovations to be more meaningful in the digital era to make agricultural extension more relevant and reliable. The innovative extension models by the Professor Jayashankar Telangana Agricultural university to bridge the gap between the research and farming community are unique. They are evolving in nature to suit better to the diversified farming.

References*

1. Agriwelfare.gov.in/en/Dept.
2. Agwu, A.E., Dimelu, m.u and Madukwe, M.C. 2008. Innovation System Approach to Agricultural Development: Policy implications for agricultural Extension delivery in Nigeria. African journal of biotechnology. 7 (11): 1604-1611.
3. All India Report on Agriculture Census. 2015-16. Department of Agriculture, Cooperation & Farmers Welfare, Ministry of Agriculture & Farmers Welfare, Government of India. 2020.
4. Ataharul Huq Chowdhury, Helen Hambly Odame and Cees Leeuwis. 2013. Transforming the Roles of a Public Extension Agency to Strengthen Innovation: Lessons from the National Agricultural Extension Project in Bangladesh. Journal of Agricultural Education and Extension. 20 (1): 1-19.
5. Gulati A, Sharma P, Samantara A and Terway P. 2018. Agriculture extension system in India: Review of current status, trends and the way forward. Indian Council for Research on International Economic Relations.
6. ICAR, Annual Report. 2023-24. Indian Council of Agricultural Research Department of Agricultural Research and Education, Ministry of Agriculture & Farmers Welfare Government of India New Delhi. www.icar.org.in.
7. Ravi Nandi and S. Nedumaran. 2019. Agriculture Extension System in India: A Meta-analysis. Research. Journal of Agricultural Sciences. 10(3): 473-479.
8. Report of the Committee on Doubling Farmers' Income Volume I "March of Agriculture since Independence and Growth Trends". Ministry of Agriculture and Farmer Welfare. (https://agriwelfare.gov.in/Documents/DFI%20Volume%201.pdf).

* The content of this chapter was prepared in consultation with above references.

18

Agricultural Extension System: Review, Trends and Way Forward at RPCAU, Pusa

M. Rai, B. Satpathy, R.K. Jha, Phool Chand and M.L. Meena

Directorate of Extension Education, RPCAU, Pusa, Bihar

bineeta.satpathy@rpcau.ac.in

Abstract

Agricultural extension systems are essential in bridging the gap between research and practical farming applications. They provide farmers with access to modern technologies, sustainable practices, and market insights to improve productivity and resource management. Dr. Rajendra Prasad Central Agricultural University's Directorate of Extension Education exemplifies this approach, offering training and guidance to farmers through workshops and consultations. A comprehensive analysis of farming systems is essential to identify training needs, address technology adoption gaps, and highlight production constraints. By addressing technology gaps, promoting sustainable farming, and connecting farmers to global markets, agricultural extension helps enhance crop yields, boost profitability, and improve rural livelihoods, ultimately supporting the goal of doubling farmers' incomes. These improvements will result in better resource management, increased crop yields, diversified practices, and improved livelihoods in rural communities.

Keywords: *Diversified Practices, Rural Communities, Sustainable Practices Training, Technology Gaps*

Introduction

Agricultural extension systems are vital in bridging the gap between research and practical application by farming communities. By providing farmers with access to the latest technologies, practices, and knowledge, these services help improve crop yields, optimize resource use, and promote sustainability. Beyond just technical advice, agricultural extension often involves helping farmers adapt to new market trends, climate changes, and government policies.

Extension agents typically work directly with farmers through meetings, one-on-one consultations, and community-based demonstrations.

Dr. Rajendra Prasad Central Agricultural University, originally known as Rajendra Agricultural University, was established as India's first Imperial Agriculture Research Institute. The foundation stone was laid by Lord Curzon on April 1, 1905, with financial support from American philanthropist Henry Phipps, Jr. Located in Pusa, Bihar. It is an Institute of National Importance. The University focuses on teaching, research, and extension, with the Directorate of Extension Education founded in 1977. This Directorate bridges the gap between technology generation and dissemination by coordinating with the State Department of Agriculture, ICAR, and other research and training institutes. It aims to deliver the latest agricultural information to farmers through its colleges and Krishi Vigyan Kendras (KVKs), as well as plan and monitor extension activities based on research results.

Public spending on Agricultural extension is one of the key policy instruments of the government to provide growth and alleviate property in rural areas. A comprehensive analysis of farming systems is essential to identify areas that need training, highlight technology gaps, address the adoption gap, and recognize production constraints. The goal is to develop and showcase alternative sustainable farming practices that enhance farmers' productivity and profitability while conserving biodiversity and improving environmental health. Additionally, farmers should be connected to the rapidly evolving global market by providing them with relevant information and guidance to help them adapt to new opportunities and challenges.

In addition to the resource and staffing limitations, the low investment in agricultural research and extension services—only 0.7% of agricultural

GDP—further compounds the problem. The World Bank's recommendation of investing 2% of Agricultural GDP in these areas emphasizes the need for a more robust and sustainable approach to agricultural development. Increased funding could enable a range of improvements, such as:

1. Expanding outreach
2. Training and capacity building
3. Research and development
4. Enhanced technology and tools.

Finally, the chapter stresses that the Indian government's goal of doubling farmers' income by 2022 cannot be realized without a strong, well-functioning agricultural extension system. If these challenges are not addressed then the ambition to improve farmers' livelihoods will remain a difficult task.

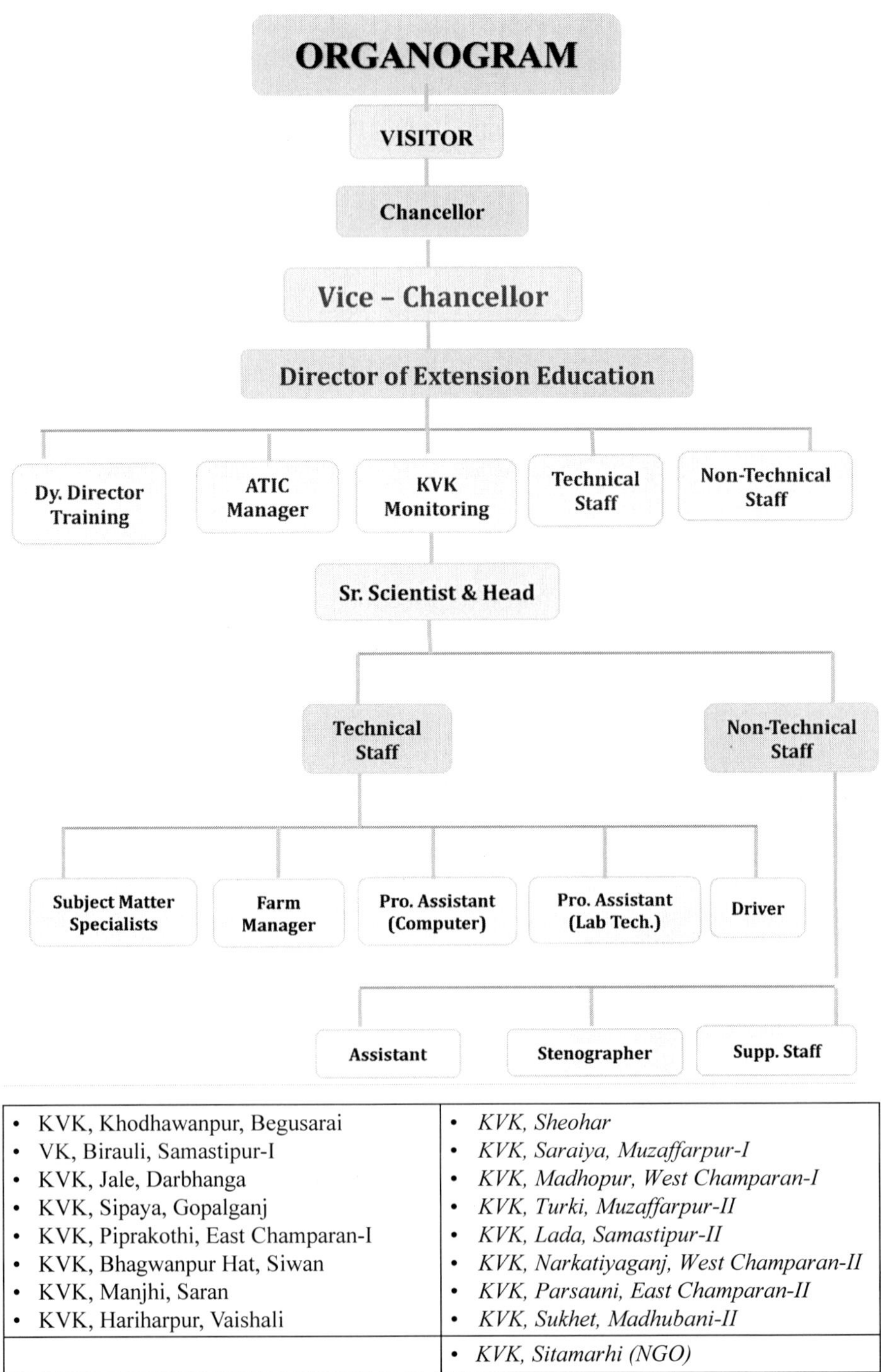

• KVK, Khodhawanpur, Begusarai • VK, Birauli, Samastipur-I • KVK, Jale, Darbhanga • KVK, Sipaya, Gopalganj • KVK, Piprakothi, East Champaran-I • KVK, Bhagwanpur Hat, Siwan • KVK, Manjhi, Saran • KVK, Hariharpur, Vaishali	• *KVK, Sheohar* • *KVK, Saraiya, Muzaffarpur-I* • *KVK, Madhopur, West Champaran-I* • *KVK, Turki, Muzaffarpur-II* • *KVK, Lada, Samastipur-II* • *KVK, Narkatiyaganj, West Champaran-II* • *KVK, Parsauni, East Champaran-II* • *KVK, Sukhet, Madhubani-II*
	• *KVK, Sitamarhi (NGO)*

Capacity Development Programme

The Directorate of Extension Education plays a crucial role in strengthening the agricultural extension system by providing need-based training to both extension functionaries and farmers. These training programs are designed to equip participants with updated knowledge and practical skills that are essential for improving agricultural productivity and sustainability. By exposing both extension workers and farmers to the latest developments in agricultural research and techniques, the Directorate aims to help them effectively meet their needs and achieve their goals.

In 2014-15, India allocated around Rs. 18 billion for agricultural extension & training exclusively, which has grown from Rs 6.4 billion in 2000-01, this recording a CAGR of 76 percent forth given period.

Capacity development of farmers and extension functionaries is indeed a key component of agricultural extension services. It involves improving the skills, knowledge, and capabilities of both farmers and those responsible for delivering extension services (extension agents or functionaries) to enable them to perform effectively and improve agricultural productivity.

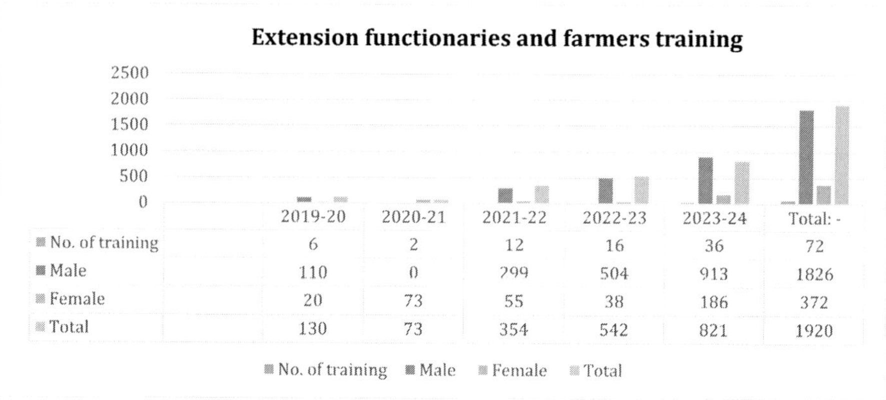

	2019-20	2020-21	2021-22	2022-23	2023-24	Total: -
No. of training	6	2	12	16	36	72
Male	110	0	299	504	913	1826
Female	20	73	55	38	186	372
Total	130	73	354	542	821	1920

Fig 1. Extension functionaries and farmers training

Training of Practicing Farmers/Farm Women

Krishi Vigyan Kendra (KVK) is an integral part of the National Agricultural Research System (NARS), aims at assessment of location-specific technology module in Agriculture & allied enterprises through technology assessment, refinement and demonstration. KVKs have been operating as knowledge Resource Centre (KRCS) of technology supporting initiative of public, private & voluntary sector for improving the agriculture economy & are connecting the NARS with extension system & farmers. Que of the prime mandate is to apart from many is capacity development of farmers and extension personnel to update then knowledge and skills on advanced technology. RPCAU Extension System adopts a multifaceted approach to identify farmers training needs effectively.

Through targeted training programs, KVK helps bridge the knowledge gap, ensuring that agricultural practices remain efficient, sustainable, and economically viable. Training provides an opportunity to sharpen the skills and knowledge of the practicing farmers and farm women. During the last 5 years, KVKs organized various training programs on the basis of the needs and interest of the farmers, which is as follows:

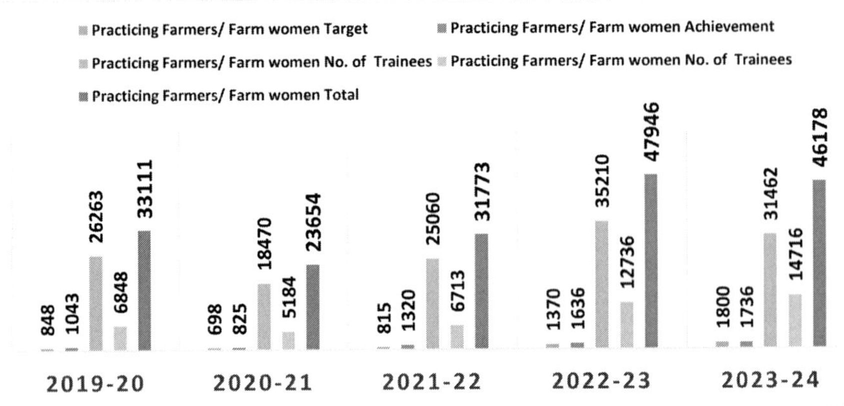

Fig 2. Practicing Farmers/ Farm women

Training of Rural Youth

Skill training of Rural Youth is a flagship scheme of the Government. This skill-based training is aimed at impacting skills to rural youth agri-based vocational areas to promote rural employment and also assist in creating skilled manpower to perform farm and off-farm activities sparring farm agriculture, horticulture, apiculture, sericulture, dairy, floriculture and fisheries.

Farmers are at the core of the agricultural value chain, and improving their capacity can directly enhance farm productivity, sustainability, and income. This involved skill development. Training farmers in modern farming techniques, such as integrated pest management, organic farming, and precision agriculture, can lead to better crop management and higher yields. As of December 2024, India's unemployment rate decreased to a three-month low of 8.65%, down from 8.88% in November, primarily due to a significant reduction in rural unemployment.

Unemployment is the biggest problem for the rural youth. They have limited access to quality education for developing their skills. When they are empowered with skills and knowledge, they are set to progress by translating their ideas, innovations for development not only for their employment but they also generate employment for others. With these objectives Directorate and KVKs train the rural youth below is a glimpse of the same:

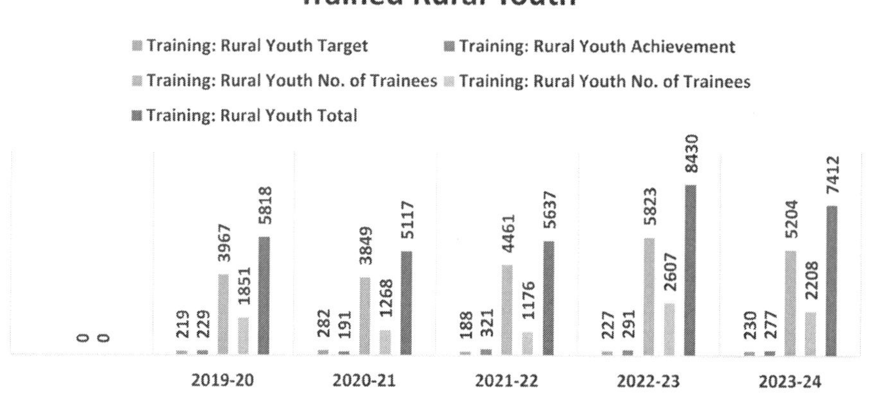

Fig 3. Trained Rural Youth

Training of Extension Functionaries

Training Extension functionaries in agriculture is carnival for effectiveness agriculture development and organization. The KVKs act as agricultural knowledge centres in the district. With the limited human resources available at KVK it is beyond the capacity to impart the skill development training and updated knowledge for agricultural development for the whole district. The state government has a well knitted network for agricultural extension at grassroot level. This grassroot level agriculture extension worker need to be trained with updated knowledge and skill for the agricultural development at village level. With these objectives, grassroot level extension worker is being trained at KVKs and Directorate.

Detail of the achievements made by the Directorate and KVKs in this endeavour to trained extension functionaries are given below:

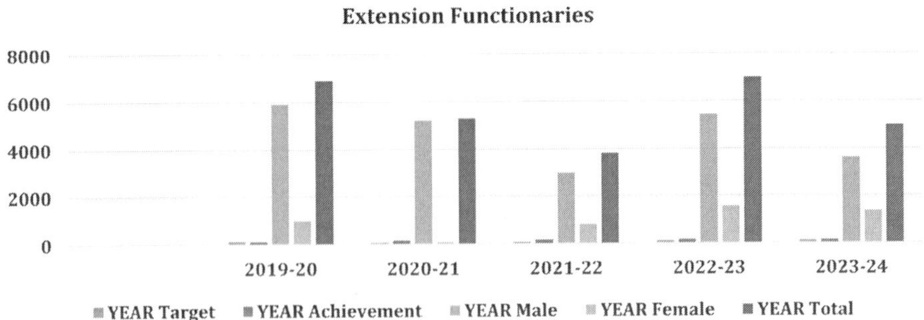

Fig 4. Extension Functionaries

Sponsored Training Programme

Agriculture employs about 42% of India's workforce, or more than 300 million people. Globally, the sector employs around 26% of workers, with significant variation between regions. Different organizations are working to use this potentiality for creation of employment among the stakeholders. For this purpose, the different organizations plan different training programs for development of skills for creation of entrepreneurship among the stakeholders. In this line, Directorate extends the support for creation of entrepreneurship. The sponsoring agencies use the expertise of the University for Development of skill for their stakeholders through skill development training program under different domains. Detailed account of sponsored training program are depicted graphically below:

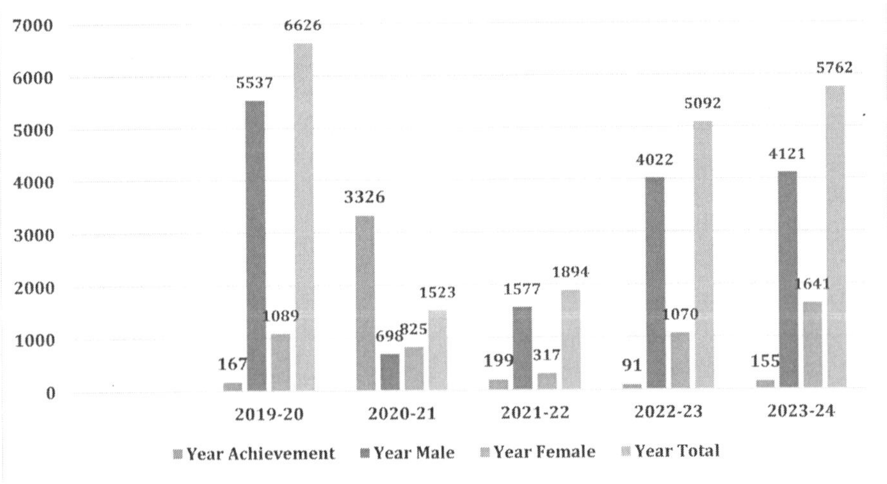

Fig 5. Sponsored Training Programme

Stakeholders Convergence approach is a recurring approval to enterprise development. The sponsored program presents the dynamics of sustainable learning in the process of supply chain.

Front Line Demonstration

Systemic creation of local proof of both the application and profitability of the recommended technology with the cooperation and participation of the farmers and under the personal guidance of the scientists and/or extension personnel is of utmost important. The primary goal of FLD is to demonstrate the effectiveness potential of new technologies and practices to farmers and extension functionaries. FLDs preside direct link belmein research & farmers. They demonstrate newly released verities, management practices & other innovations. The perceived benefits of FLDs are improved technology adoption, enhanced research and intimate knowledge sharing.

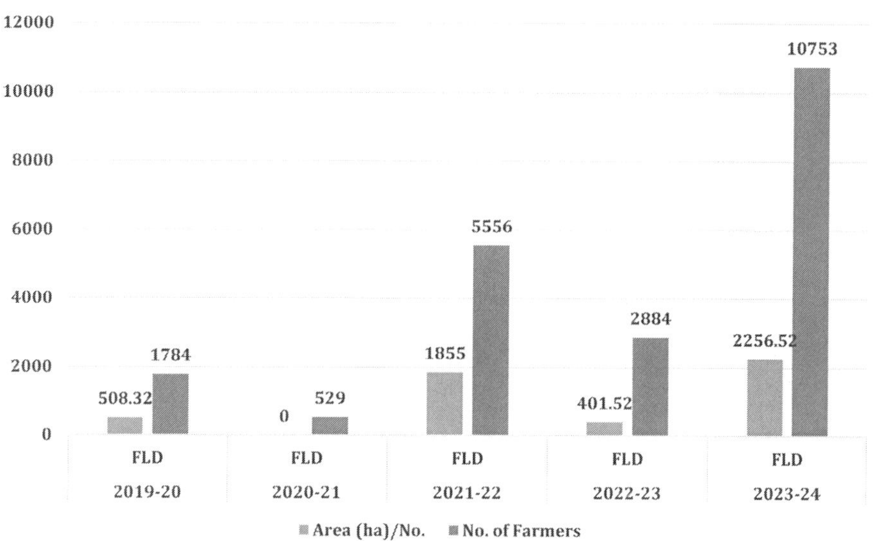

Fig 6. Front Line Demonstrations

Technology Demonstration for its Application

Cluster front time demonstrations are a unique approach by the ICAR to showcase improved technologies and practices to farmers with scientist directly involved in planning, extension and monitoring to enhance crop production & productivity and provide feedback for further research.

Demonstration of CFLD (Pulses & Oilseed) Technologies at the Farmers' field

Cluster frontline demonstration is a unique approach by the Indian Council of Agricultural Research on oilseed and pulse crops to provide a direct interface between scientists and farmers where farmers are guided by the KVK Scientists during demonstrations for implementation of improved technologies like seed treatment, IPM, INM, land preparation, etc. CFLD project started since 2015-16 under National Food Security Mission (NFSM) KVKs conducted Cluster Frontline Demonstration (CFLD) to demonstrate the production potential of newly released technologies on the farmers' fields at different location in a given farming system, and organized farming and extension activities for farmer and extension workers for dissemination of various technologies.

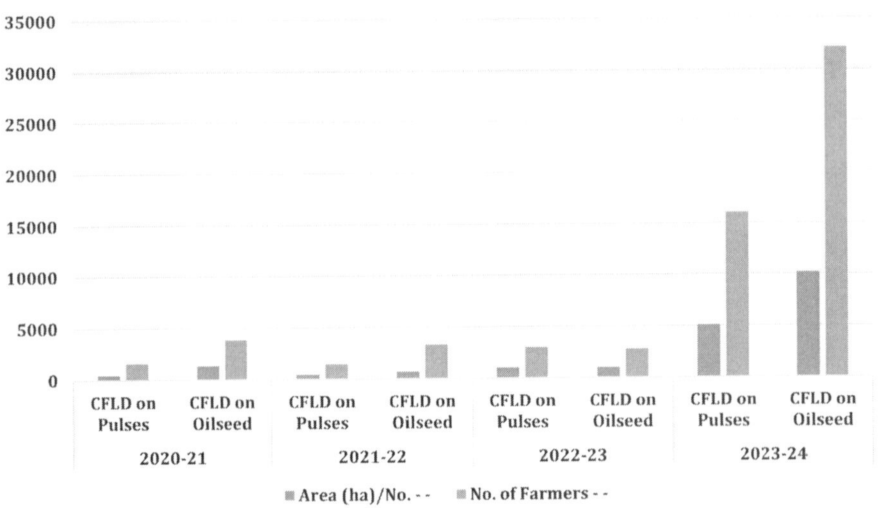

Fig 7. Technology Demonstration for its application

In Bihar, where oilseeds production is a deficit, the state cultivates several oilseed and pulses, including grant giant, rapeseed, mustard and arhar, urad, moong and Gram.

On Farm Trial (OFT)

On farm trial is an approach of adaptive research conducted on farmer's field within their farming system perspective with their active participation and under their management. The objective of OFT is to develop technologies which might help solve problems of group of farmers in a defined study area. In last five years all sixteen KVKs have assessed 126 technologies. During 2022-23, around 165 technologies were assessed.

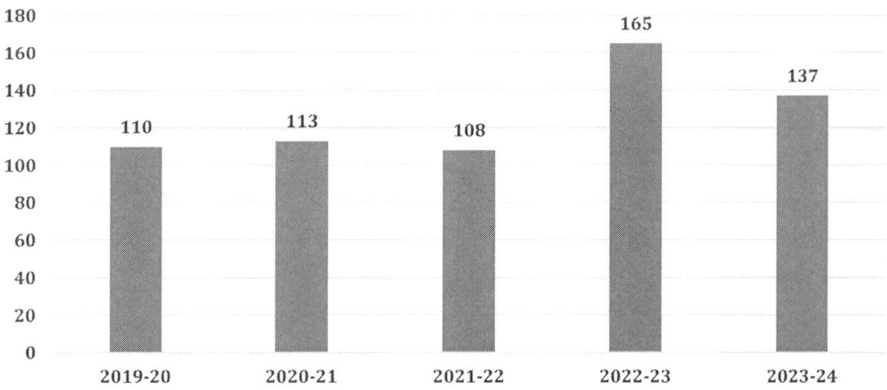

Fig 8. On Farm Trial

Seed Production

The quality of seeds is considered as an important factor for increasing yield. The use of quality seeds helps greatly in higher production/ unit area to attain food security of the country. Quality seeds have the ability for efficient utilization of the inputs such as fertilizers and irrigation. Quality seed production is one of the important mandates of KVKs and they produced maximum seeds i.e. 6247.84 qt. of cereals, pulses oilseeds, sugarcane, potato, etc., in the year 2022-23.

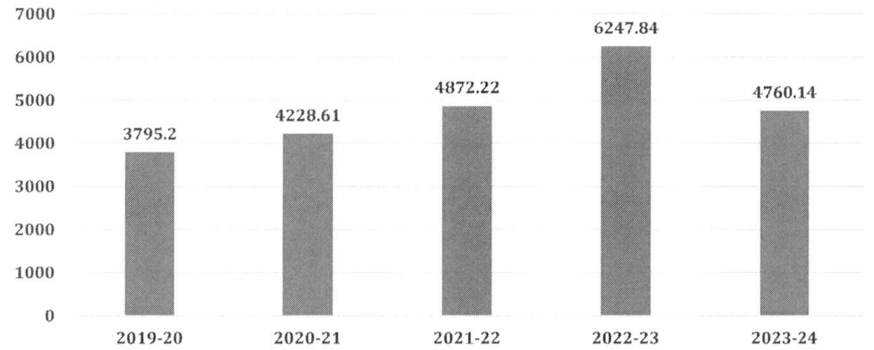

Fig 9. Seed Production (qt.)

Fruits, Vegetable, Flower and Other Planting Material

Important objective of Krishi Vigyan Kendra is to produce uniform, healthy, disease-free planting material raised through seed or vegetatives with an overall goal to raise the physiological and phytosanitary quality of the plant available to stakeholders and to increase productivity.

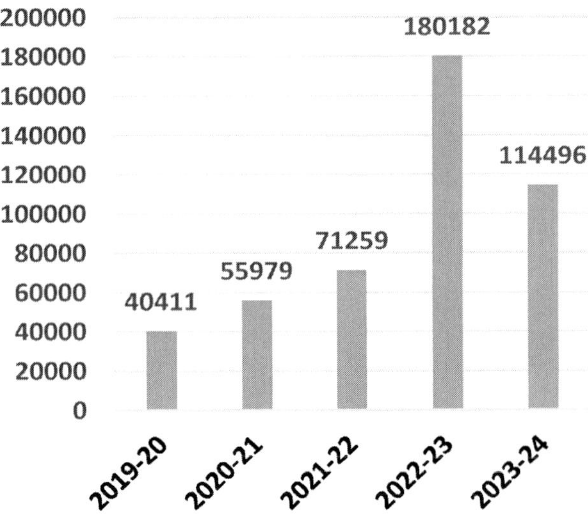

Fig 10. Fruits Planting Material (Nos.)

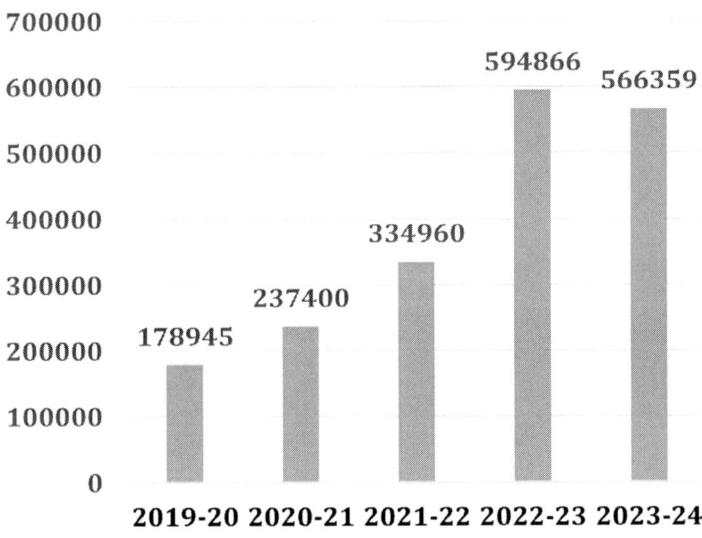

Fig 11. Vegetable, Flower and Other Material (Nos.) produced

Fig 12. KVK, Piprakothi received 1st Prize by ICAR Pandit Deendayal Upadhyay Krishi Vigyan Protshan Puraskar-2020

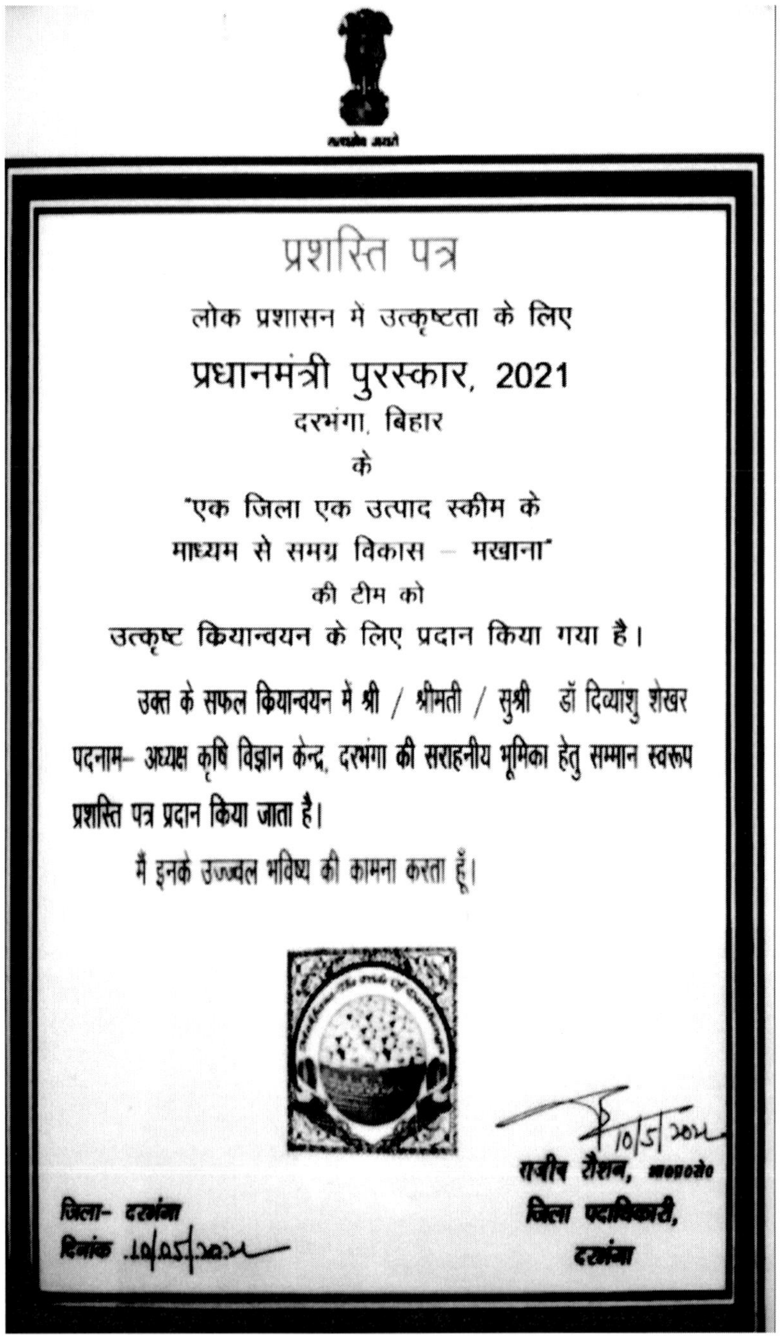

सत्यमेव जयते

प्रशस्ति पत्र

लोक प्रशासन मे उत्कृष्टता के लिए

प्रधानमंत्री पुरस्कार, 2021

दरभंगा, बिहार

के

"एक जिला एक उत्पाद स्कीम के
माध्यम से समग्र विकास – मखाना"

की टीम को

उत्कृष्ट कियान्वयन के लिए प्रदान किया गया है।

उक्त के सफल कियान्वयन में श्री / श्रीमती / सुश्री डॉ दिव्यांशु शेखर पदनाम- अध्यक्ष कृषि विज्ञान केन्द्र, दरभंगा की सराहनीय भूमिका हेतु सम्मान स्वरूप प्रशस्ति पत्र प्रदान किया जाता है।

मैं इनके उज्जवल भविष्य की कामना करता हूँ।

10/5/2022

राजीव रौशन, भा०प्र०से०
जिला पदाधिकारी,
दरभंगा

जिला- दरभंगा
दिनांक 10/05/2022

Fig 13. Certificate of appreciation awarded by Hon'ble Prime Minister to Krishi Vigyan Kendra, Jale for holistic development of ODOP-Makhana

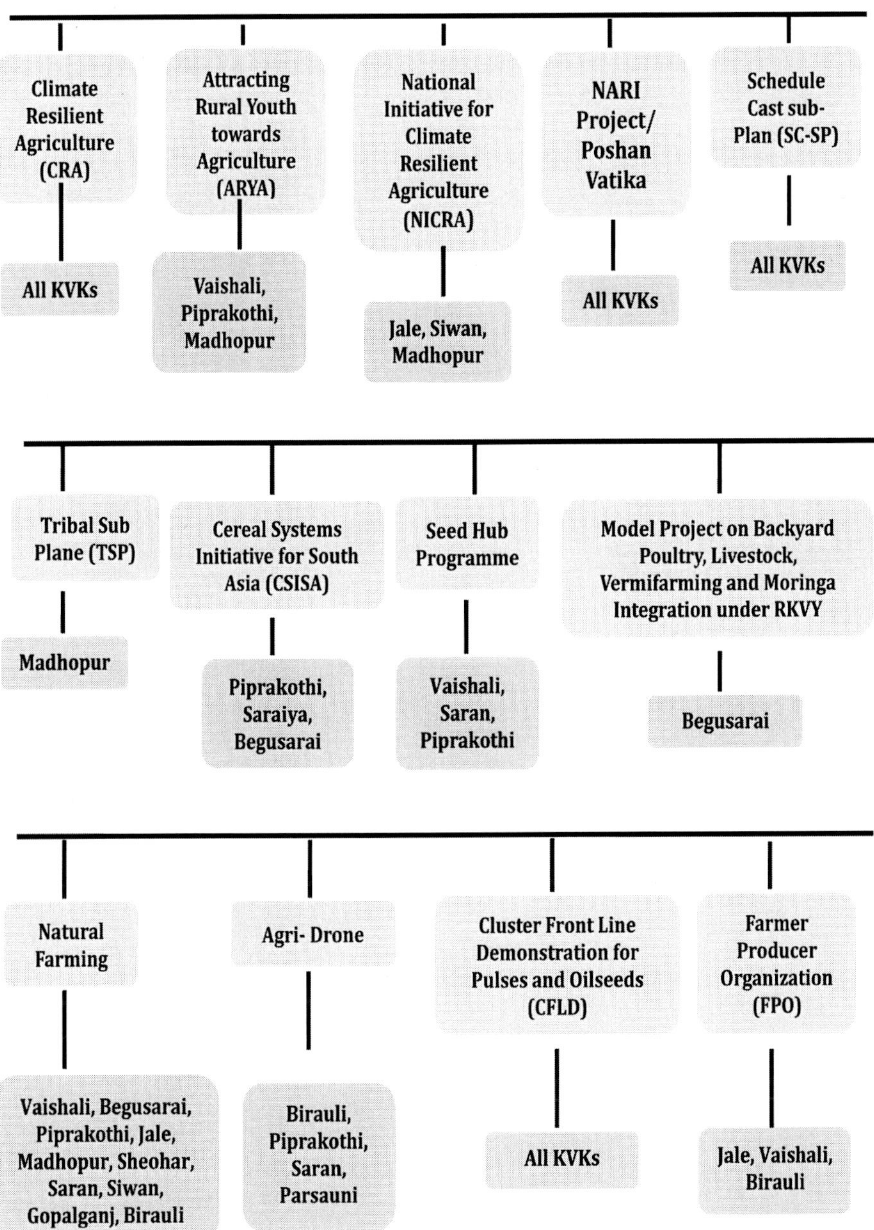

Fig 14. Projects running under Krishi Vigyan Kendra

Fig 15. 29th August 2021 Sukhet Model - Monetizing of Cow Dung

Fig 16. February, 2019 Smt. Rajkumari Devi, Hon'ble Prime Minister Shri Narendra Modi Ji expressed his views about Women empowerment by citing the example of Kisan Chachi in his 53rd Edition of MANN KI BAAT

Fig 17. Smt. Manorma Singh, woman-entrepreneur on mushroom promoted by KVK, Vaishali won received "Jagjivan Ram Abhinav Krishi Puraskar-2020".

Fig 18. Sri Shiv Prasad Sahani, Siwan and Dr. R. S. Singh, Saraiya won Jagjivan Ram Abhinav Kisan Puruskar 2020 (Zonal) Award by ICAR New Delhi

Fig 19. Sri Jitendra Singh, Vill-Namidih and Md. Musharraff Khalif, Bakhari Faridpur promoted by KVK, Vaishali awarded with IARI fellow Award and IARI Innovative Farmers Award respectively

Fig 20. Sri Shiv Prasad Sahani, Siwan got "IARI Fellow Farmer" 2023 by IARI New Delhi

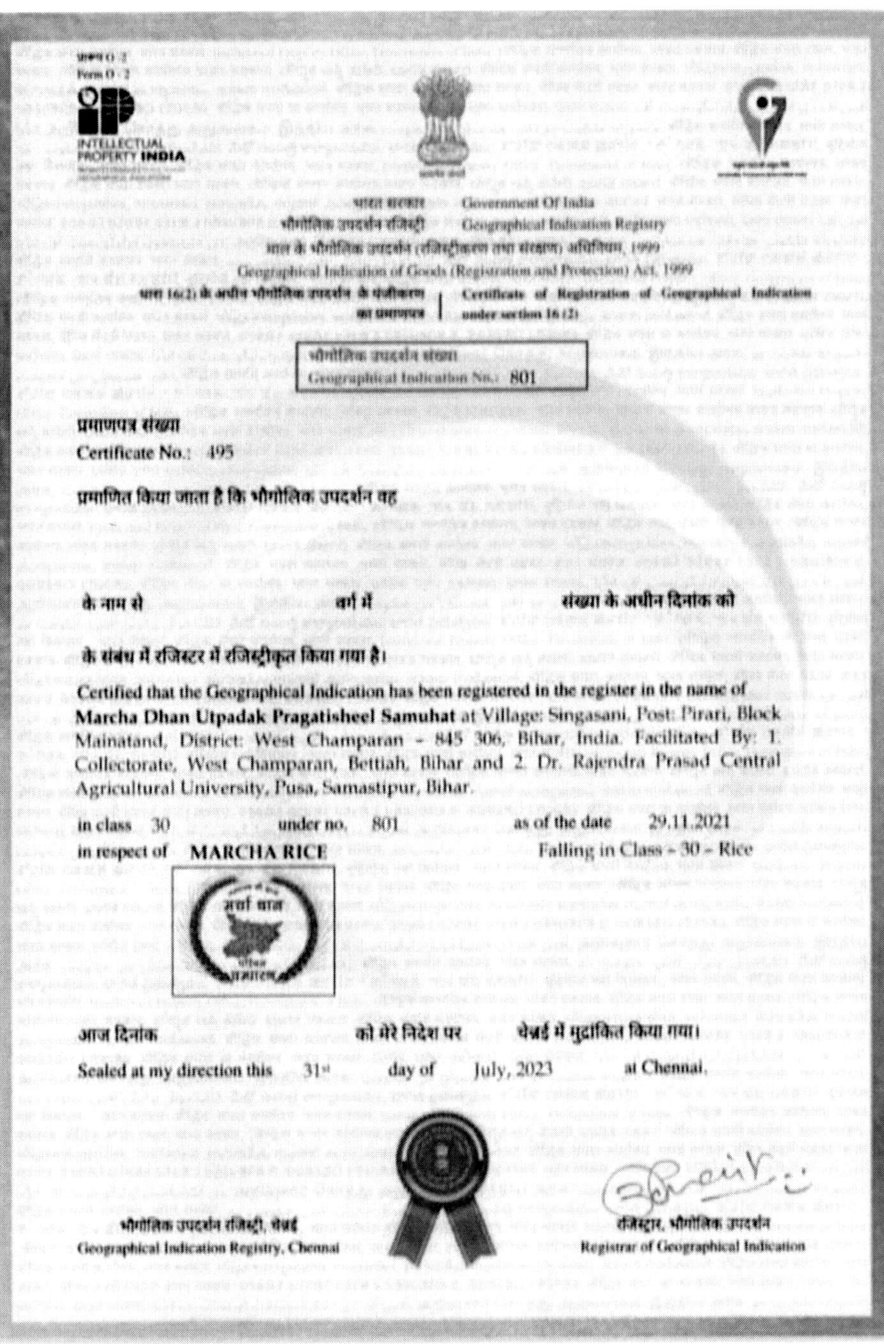

प्रपत्र G-2
Form G-2

INTELLECTUAL PROPERTY INDIA

भारत सरकार | Government Of India
भौगोलिक उपदर्शन रजिस्ट्री | Geographical Indication Registry
माल के भौगोलिक उपदर्शन (रजिस्ट्रीकरण तथा संरक्षण) अधिनियम, 1999
Geographical Indication of Goods (Registration and Protection) Act, 1999

धारा 16(2) के अधीन भौगोलिक उपदर्शन के पंजीकरण का प्रमाणपत्र | Certificate of Registration of Geographical Indication under section 16 (2)

भौगोलिक उपदर्शन संख्या
Geographical Indication No.: 801

प्रमाणपत्र संख्या
Certificate No.: 495

प्रमाणित किया जाता है कि भौगोलिक उपदर्शन यह

के नाम से वर्ग में संख्या के अधीन दिनांक को

के संबंध में रजिस्टर में रजिस्ट्रीकृत किया गया है।

Certified that the Geographical Indication has been registered in the register in the name of **Marcha Dhan Utpadak Pragatisheel Samuhat** at Village: Singasani, Post: Pirari, Block Mainatand, District: West Champaran - 845 306, Bihar, India. Facilitated By: 1. Collectorate, West Champaran, Bettiah, Bihar and 2. Dr. Rajendra Prasad Central Agricultural University, Pusa, Samastipur, Bihar.

in class 30 under no. 801 as of the date 29.11.2021
in respect of **MARCHA RICE** Falling in Class - 30 - Rice

आज दिनांक को मेरे निदेश पर चेन्नई में मुद्रांकित किया गया।
Sealed at my direction this 31st day of July, 2023 at Chennai.

भौगोलिक उपदर्शन रजिस्ट्री, चेन्नई
Geographical Indication Registry, Chennai

रजिस्ट्रार, भौगोलिक उपदर्शन
Registrar of Geographical Indication

Fig 21. Geographical Indication for Marcha Dhan

Salient Achievements of KVKs

1. **The University Implementing Climate Resilient Agriculture (CRA)** Programme funded by the Govt. of Bihar to show the efficacy of the climate resilient technologies in the farmers' field by all 16 (sixteen) KVKs of RPCAU, Pusa. Every KVK is working in a cluster of five villages for demonstration of the Climate Resilient Technologies in the farmers' field. This Climate Resilient Technology includes Zero-tillage cultivation of wheat, mustard and lentil during *Rabi* season and DSR, raised bed technology for cultivation of pigeon pea and maize etc., during *Kharif* season along with other climate resilient technologies. In these activities, the salient achievements have been observed in cultivation of pigeon pea in bund area and the yield of pigeon pea was more than the traditional cultivation.

2. **Rejuvenation of the Technologies:** Sugarcane is the main cash crop in the district of East and West Champaran. In the traditional way, seed rate for cultivating sugarcane is 25-30 quintal/acre. Settling Transplantation Technology (STT) drastically reduces seed requirement of sugarcane

which is 6-8 quintal per acre. In this technique, seedling need to be raised through portray. It helps in production of disease-free planting materials. This technique was developed by ICAR-IISR, Lucknow. KVK, Narkatiaganj with little modification of the original technique has standardized the technique and popularized it in the Narkatiaganj area. During 2021-22, this technique was demonstrated in the farmers' field which was adopted in 100 acres of land benefitting several farmers.

3. **Integrated Farming System (IFS):** This system is self- sustainable and very much effective for the farmers where output of one system used as input for other system. In this process, nine KVKs of RPCAU have developed integrated farming model to impart the training to the farmers. The details of the integrated farming model have been depicted at the figure. The economics of the integrated farming system is presented in the Table.

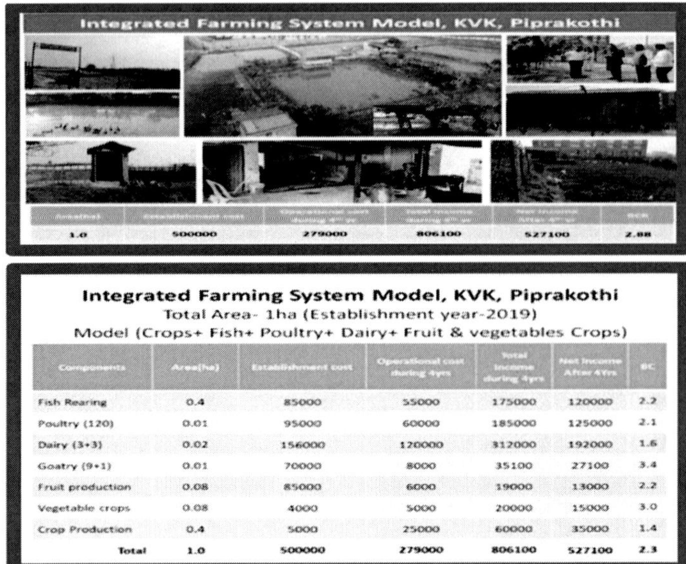

Integrated Farming System Model, KVK, Piprakothi
Total Area- 1ha (Establishment year-2019)
Model (Crops+ Fish+ Poultry+ Dairy+ Fruit & vegetables Crops)

Components	Area(ha)	Establishment cost	Operational cost during 4yrs	Total Income during 4yrs	Net Income After 4Yrs	BC
Fish Rearing	0.4	85000	55000	175000	120000	2.2
Poultry (120)	0.01	95000	60000	185000	125000	2.1
Dairy (3+3)	0.02	156000	120000	312000	192000	1.6
Goatry (9+1)	0.01	70000	8000	35100	27100	3.4
Fruit production	0.08	85000	6000	19000	13000	2.2
Vegetable crops	0.08	4000	5000	20000	15000	3.0
Crop Production	0.4	5000	25000	60000	35000	1.4
Total	1.0	500000	279000	806100	527100	2.3

4. **Secondary Agriculture**

   a) The KVKs of RPCAU are engaged in popularization of the secondary agriculture under creation of wealth from waste. In this endeavor, the village waste is being collected and made compost for using it in the farmers' field and in lieu of that the farmers used to get one cylinder in every two months intervals those who are associated in this project. This model created the huge publicity. As a result, it appears in the "Pradhan Mantri Maan Ki Baat Programme."

b) Empowering women with Secondary Agriculture: Secondary agriculture was tried in one of the KVKs by making "Aachar (Pickle)" by using mango and other agricultural produce. Smt. Rajkumari Devi undergone training at KVK, Saraiya. Later she formed self-help group with 300 women for preparation of Aachar to earn their livelihood by using this technique. This has also appeared in the "Pradhan Mantri Maan Ki Baat Programme".

c) Banana is the most prevailing fruit crop produced in Bihar. After harvesting banana, pseudo-stems remain un-used. The KVK used the pseudo stems for extracting fibres which is being used for different products. At present, five farmers have developed banana fibre extraction units and earning.

d) Mushroom Cultivation is another secondary agriculture which is created euphoria among the farmers mostly women farmers. The women farmers are also making different value-added products from the Mushroom. As a result Bihar became the highest producer of Mushroom in the country.

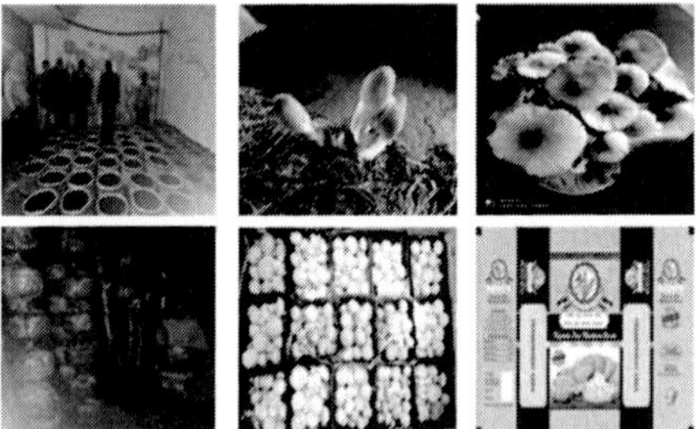

5. Innovative Technology adopted at KVKs

a) The KVKs of RPCAU are situated in the flood prone areas of Bihar. During the rainy season, most of the lands get inundated and not suitable for cultivation of the vegetable so the cost of the vegetable during the rainy season crossed many folds. Gunny bags technique was tried in the Rice field for cultivation of the vegetable especially for the creepers and gave good response. While the field remain submerged.

b) The Litchi and Mango are prevailing plantation crops in this region. KVKs tried for diversification of fruit crops by introduction of sweet lime and mandarin orange.

c) Development of the livestock, poultry and vermi-compost as a self-sustainable model: Infertility in cow became a problem in the village. The farmers let loose the infertile cows in the street. It created a menace in the society, keeping this in view, our Begusarai, KVK collected 11 infertile cows from the street and started feeding with introduction of scientific feeding and management. After a year of rearing, these three cows got pregnant and deliver the calves and started producing milk.

6. Infrastructure Development

a) This University got sanctioned of five (05) new KVKs. During this period, all the 5 KVKs have completed the construction of Administrative Building as well as Farmers' Hostel. The Administrative Building and Farmers' Hostel have been inaugurated by the Hon'ble Minister of Agriculture & Farmers Welfare, Govt. of India on 28th February, 2023.

b) All the KVKs have procured different agricultural implements for farming operation. To keep these agricultural implements, the implement shed were constructed at 12 (twelve) KVKs which are being used for dual purposes.

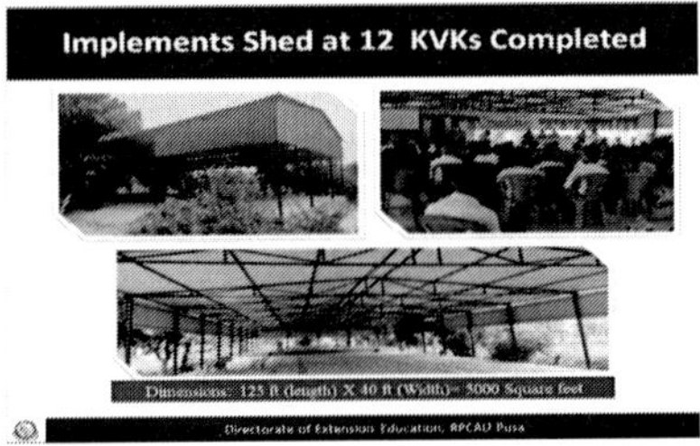

An Initiative of the Government of Bihar under 4th Krishi Road Map

Fig 22. ICT- enabled "KRISHI GYAN VAHAN" for dissemination of Agricultural technology and information

Mandates of Krishi Gyan Vahan

- Collection of samples for soil testing directly from farmers' fields
- Solution to agriculture related problems at the doorstep of farmers
- Solution of practical problems related to agriculture and allied sectors by having live contact with the farmers
- Solutions related to the health of animals and birds along with investigation of insect diseases of food grains/garden/other crops
- Provision of technical films for enhancing the knowledge of farmers

- Awareness and capacity building of farmers through virtual medium by the University
- To motivate rural youth for self-employment
- Agricultural inputs like seeds, organic manure, liquid bio-fertilizer along with mushroom spawn will be available on demand
- To make farmers aware about public welfare government schemes (small marginal women SC-SP)
- Awakening entrepreneurship ecosystem.

The progress and emerging trends provide a comprehensive approach to improving India's agricultural extension system. They address the critical issues while providing clear, actionable solutions as follows :

1. Enhanced Investment

Action: Targeting 2 percent of agricultural GDP for funding, as suggested by the World Bank, would ensure that both agricultural research and extension services have the resources they need to function optimally. This investment would not only increase the capacity of extension services but also fuel research and development in agriculture, leading to more innovative solutions for farmers.

2. Expansion of Extension Services

Action: Extending the scope of extension services to cover not just crops but also livestock, fisheries, and horticulture would offer a more holistic approach to rural development. Different farming sectors often require specialized advice and support, and integrating these sectors ensures that no farmer is left behind.

3. Adequate Training and Hiring of Personnel

Action: Hiring more extension professionals to meet the recommended 1:750 extension worker : farmer ratio and ensuring proper training is essential. Skilled extension agents are key to the successful implementation of the extension services. Regular training would ensure that agents remain updated on the latest agricultural technologies and methods.

4. Reducing Non-Extension Duties

Action: Shifting non-extension responsibilities, such as input distribution and subsidy management, away from the extension services would help them refocus on their core function—providing farmers with expert guidance and knowledge. This would also reduce the workload on extension staff, allowing them to deliver more effective services.

5. Leveraging Technology

Action: Integrating digital platforms, mobile-based solutions, and other technologies into the extension system can help reach more farmers, especially those in remote areas. Technologies like mobile apps, online advisory services, and even drones for precision farming could improve how information is delivered.

Conclusion

By implementing these recommendations, agricultural extension system could become a powerful tool to support farmers, enhance agricultural productivity, and ultimately help achieve the government's goal of doubling the farmers' incomes. These changes could lead to better resource management, increased crop yields, more diversified agricultural practices, and improved livelihoods in rural communities.

19

Innovative Extension Approaches of Assam Agricultural University for Enhancing Farmers' Income

C.K. Sarma[1], M. Neog[2] and Bidyut C. Deka[3]

[1]Chief Scientist, All India Coordinated Research Project on Irrigation Water Management, AAU, [2]Director of Extension Education, AAU, [3]Vice Chancellor, AAU, Jorhat-785013

Email: manoranjan.neog@aau.ac.in

Abstract

Agricultural extension system plays a very crucial role to bring desired changes to face the present-day challenges in agricultural sector. Production-led strategy which was the primary focus of extension system earlier, needs to be expanded to embrace a market-led strategy to deal with the new challenges. So, a transformation is required from technology-driven extension system to market-driven system where farmer or farmer groups are able to market high value products, maintain quality of produces and achieve market demands. Now, consumer needs diverse, safe and nutritious food which calls for a major reorganisation of the agricultural value chains to support transformation from the traditional production-oriented supply chains to demand based approaches. Innovative extension approach must be worked on how to make agriculture more profitable to provide livelihood security to farmers. But to play this role effectively, extension should expand its mandate beyond disseminating information on technologies so that it can better respond to the evolving demands for support and services of farmers. Assam Agricultural University launched various innovative outreach programs so as to create a direct interface between scientists/ faculties and farmers with the primary goal to ensure reaching of modern agricultural research and extension services to rural communities more effectively, empowering them with the knowledge and tools to increase productivity, income, and resilience. As a result of these programs, there has been mobilization of intellectual resources of the University for village-level development, promoting stronger scientist-farmer relationships and building a support framework for sustainable agricultural growth

Keywords: *Extension system, farmers' income, innovative approaches*

Introduction

The agriculture scenario has evolved considerably over the years. In the present context, consumers need diverse, safe and nutritious food which calls for a major reorganisation of the agricultural value chains to support transformation from the traditional production-oriented supply chains to demand-based approaches. Trade liberalization has further intensified competition by exposing farmers to both domestic and international markets, compelling them to meet elevated standards in terms of product quality and pricing. Therefore, the primary focus should be on enhancing technical expertise, managerial capabilities and access to timely information to enable smallholder farmers not only to strengthen their livelihoods but also enhance their marketing skills amid the fast-evolving global economic landscape.

Access to true information is equally essential for addressing growing challenges related to natural resource degradation and climate change. Extension in Indian context, play a vital role in bridging the knowledge gap between research institutions and farmers by providing a range of advisory amenities and facilitate expertise application, transmission and management in the farmer's fields. Perhaps, agricultural extension is the only service at the ground-level that can provide essential knowledge, guidance, information as well as assistance to the farmers and the stakeholders. There has been extensive range of transformation and inventiveness in agricultural extension services in India; however, the exposure and access to information which were disseminated to marginal and underprivileged farmers are uneven and patchy. Likewise, in real situation a huge gap existed between the technology developed/released and the technology disseminated/adopted.

A study by NSSO (National Sample Survey Organisation), 2014, highlighted that the almost 59 per cent of the farm families refrained from any assistance from either government or private extension services. In the country, 40.6 per cent households received extension assistance while, only 11 per cent of the services came from physical government machineries. At present hardly one-third of the technology developed at the research organizations has reached the farming community and the statistics of adoption/application of such technology at field level is even less than that. Practical application of many of our recommended agricultural technologies in the field is very much difficult either because of the complex nature of the technology or inputs so prescribed are not readily available in nearby areas. Moreover, most of the extension machineries tend to work in isolation from each other.

The production-oriented strategy which was the sole focus of earlier extension services, needs to be expanded to market-led strategy to deal with the novel challenges. Innovative and ground-breaking extension approach must be worked on to make agriculture more profitable and sustainable imparting livelihood security to the farmers. In this new era, transformation should be aligned with technology driven extension system to market driven system. Therefore, our agricultural extension system should expand its mandate beyond disseminating technological information to farmers and should respond to the evolving demands of farmers for support and services. These include organizing user/producer groups, better linkages among input and output market communities, storage, processing, entrepreneurship development, natural resource management, and issues related to women. Our agricultural extension system should also involve in technology selection and also in research planning.

Emerging Issues and Technological Interventions

In Assam, agriculture remains the largest sector as a source of livelihood, generating employment opportunities to more than 53 per cent of the total workforce group directly or indirectly and therefore, development in agriculture and allied sector will not only decide the condition of low-income group in the state but also state's poverty condition. Efforts have been made to improve the farmers' income, however, it's been proved that farming is becoming non-profit over a period of time. It is demotivating the farmers of younger age group which is finally resulting to search alternative employment opportunities instead of farming.

Participation and retaining youth in agricultural sector can provide solution to the problem of unemployment in our country. But the question is how youth are motivated to take up agriculture as a profession and whether the agriculture sector has enough prospects to support a respectable livelihood. In India, 33.7 per cent of operational land holders belong to 41–50 years' age group while 33.2 per cent is within 51–60 years out of almost 100 million farmers. These groups of people are reaching the age of retirement from agriculture while next generation does not want to engage themselves in agriculture (Mahapatra, 2020). Many a times, traditional form of agricultural system diverts interest of the younger section from farming to other nonfarm activities.

Agriculture in Assam is mainly characterized by mono-cropping with low input as well as low output, subsistence type of farming, practised prominently under rain-fed condition. Till now, the net and gross cropped areas in the state are 28.11 (35.1 per cent of geographical area) and 40.99 lakh hectares, respectively with a cropping intensity of around 145 per cent. Improvement in productivity of crops and other enterprises is one of the most important factors that can increase farmers' income. Output per hectare, which is a common measure/scale of agricultural productivity, remains low in the state for most of the crops while compared with national averages. This is attributed to low and faulty input uses, lesser access to modern technology and no sound and real technological breakthrough in recent times.

Out of total rice (*Oryza sativa* L.) growing area (24 lakh ha) in Assam, 18 lakh ha area is under *kharif* rice production system. Although, *kharif* rice is grown in major portion of the rice growing areas in Assam, a major portion of this area remains fallow during *rabi* season of the year. In India, approximately 11.69 m ha of land is kept fallow during *rabi* and summer after harvesting of *kharif* rice. About 82 per cent of the total rice fallow area is mostly distributed in Chhattisgarh, Jharkhand, Upper Assam, Bihar, eastern Uttar Pradesh, Odisha and West Bengal (Annual Report DPD, 2026-17). With low input requirement, pulses and oilseeds may be perspective crops in rice fallows under rainfed situation. Pulses are in fact soil building crops capable of transforming our dominant cereal-based systems to an ideal and sustainable system in time to come (Praharaj and Blaise, 2016). This rice fallows establishes decent scope for crop diversification and their right utilization can help in overcoming many social and economic glitches including unemployment, poverty, low income.

Like rest of the country, Assam is extremely vulnerable to climate change as the state has high reliance on agriculture that is likely to only increase because of its growing population. Assam receives high rainfall; which is highly unpredictable – sometimes excess, sometimes less, sometimes early and sometimes late and farmers' often experience aberrant weather conditions like flood or drought like situation during crop growing period. Historically, people have developed ways by themselves to cope with climate variability which has become embedded in their lifestyles and traditions. However, traditional coping and adaptation practices developed in response to normal climatic variations are becoming less-effective against the increasing scale and intensity of the changing hazards. To address these issues, the role of extension machineries is very important.

Small and marginal farmers play crucial role in Indian economy by contributing 51 per cent of total agricultural output with 46 per cent of operational land holding and there has been continuous decline in operational land holding size in India with each successive generation. This situation has raised serious concern on the survivability of small and marginal land holders. Out of India's total farmers, 67.1% is marginal and 17.9% is small as per the Agriculture Census 2010-11 who have relatively low income than their consumption expenditure. Similarly, out of total 27.2 lakh farm families in Assam, 67.31 % are marginal and 18.25% are small. Today, small and marginal farmers are facing various challenges and in most of the cases, they are just surviving in a situation where they have no or very limited access to new technology, credit, input, farm machineries, market etc. The main reason of suffering of farm communities is realised due to the existing gap between scientific know-how and field level do-how. Now, it becomes imperative to enable our farmers with access to improved technologies, required amenities and a greater number of markets to produce higher and better-quality commodities. These necessitate collective efforts of small, marginal and landless farmers in the form of Farmers' Producers Organization (FPO) to enhance economic strength & market accessibility of farmers for raising their income.

The advancement in information and communication technology sectors in recent years provides a feasible alternative to overcome existing physical barriers in extension system. In recent years, the growth of such technologies helps to disseminate and spread the information on the most suitable and required package of practices on different aspects of sustainable agricultural development. Therefore, the use of IT application in agriculture will bring breakthrough change in the acceptance, attitude and knowledge level of user.

Innovative Extension Approaches of AAU

The agriculture and its allied services act as a principal contributor in the economy of Assam sharing about 20% of the state's GDP and supports livelihood of nearly 75% of the population. The productivity of almost all the major crops in Assam, however, stands much below the national average, highlighting a critical gap between potential and performance in the sector. As a result, there is a need to capitalize on the hidden strengths of the agriculture sector in order to exploit enormous underlying potential for maximum production and profitability. In such scenario, current reforms in agriculture sector by the Government of India for self-reliant India (*Atma Nirbhar Bharat*) coupled with the state Govt.'s special drive for increasing farmers' income offers wonderful opportunities to shape the state's agrarian economy for bridging the gap between research and farming practices in order to establish a self-reliant Assam (*Atma Nirbhar Axom*). The possible gain from all these reforms and governments' stride for transformation of agriculture sector is largely dependent on the proactive actions taken by all the concerned departments and agencies of the state. The active support of the education and research institutes in delivering need-based, location-specific technologies directly to farmers is one of the major requisites for harnessing desired benefits. Unless the farmers are adequately backstopped with situation-specific, demand-driven technologies and supported with trained manpower, the state will once again miss the bus as it happened during period of green revolution. Assam Agricultural University has been recognised for its key role in the development of the agriculture sector in the North Eastern part of the country.

a) "Amar Gaon Aamar Gaurav" program for efficient extension services to rural communities

Assam Agricultural University launched an innovative outreach program as "Aamar Gaon Aamar Gaurav" to create a direct interface between scientists/ faculties and farmers with the primary goal to ensure reaching of modern agricultural research and extension services to rural communities more effectively, empowering them with the knowledge and tools to increase productivity, income, and resilience. This program is a strategic step to mobilize the intellectual resources of the University for village-level development, promoting stronger scientist-farmer relationships and building a support framework for sustainable agricultural growth. Under the methodical establishment of this outreach program, teams are constituted with 4-5 members containing multi-disciplinary scientists headed by a senior level faculty member. The selected groups identify village clusters consisting of four villages from relatively disadvantaged areas within a radius of 100 km from their work place. Different departments like Agriculture, Veterinary & AH, Fishery and organizations like Assam State Rural Livelihood Mission (ASRLM), Panchayat and even different NGOs working in the areas including KVKs help the groups in selecting those villages. The above groups establish a solid Farmer-Scientist Interface- which encourages continuous two-way communication to facilitate the exchange of ideas, innovations, resources, and feedback. This interaction supports the development and effective dissemination of appropriate technologies tailored to local needs.

Under "Aamar Gaon Aamar Gaurav" program, these multi-disciplinary group of scientist's act as a one-stop support system for farmers and potential entrepreneurs, guiding them to the right resources. By taking the advantage of the on-going schemes of the agriculture and allied departments they ensure their effective implementation within the target areas, with strong participation from local communities. Furthermore, these groups play a key role in creating awareness about the flagship programs of the governments being implemented and encourage and handhold them to reap desired benefits of the programs/ schemes. They also keep the farmers updated about the need of adopting specific agri- related activities through ICT and provide technology backstopping as per geo-spatial conditions of the village and timely disseminate information about agricultural inputs *viz.*, seed, fertilizer, chemical, agricultural machinery, weather, market, etc. to the farmers. Moreover, these programs also create awareness about the sensitive issues of national importance such as Swachh Bharat Abhiyaan, cleanliness drive, water conservation, soil health, climate change mitigation and adaptation agro-ecology and genetic resource conservation etc.

The implementation of this initiative has been projected to enhance access of the farmers to an array of climate-smart modern crop varieties, livestock/fish breeds along with production technology for adoption leading to the validation and wider implementation of context-appropriate improved technologies. One significant change might be in the strategic reorganization of existing farm enterprises into Integrated Farming Systems (IFS) models with enhanced synergy among the enterprises, resource use, employment generation and system income. By adopting a holistic approach of combining total value chain management with IFS models, households can expect a substantial and sustainable rise in income and livelihood security. Additionally, the program focuses on enhancing the skills of farmers and rural youth, equipping them with science-led innovations and entrepreneurial capabilities, thereby driving a larger transformation in the rural economy.

b) AAU as a Cluster-Based Business Organization (CBBO) for promotion of Farmers' Producers Organization (FPOs)

Assam Agricultural University has been playing a pivotal role in the process of formation and handholding of FPOs in the state to address various issues of small and marginal farmers. Major activities undertaken by Assam Agricultural University and also achievements have been discussed below.

i. CBBO-AAU was established in 2021 with a mandate to promote and nurture FPOs under the central sector scheme "Formation and Promotion of 10,000 FPOs".
ii. During the financial year 2021-22, twenty-five (25) FPOs were formed and promoted by CBBO-AAU through its local CBBO offices in different districts of Assam.
iii. During 2023-24, eighteen (18) additional FPOs were formed and promoted by CBBO-AAU.
iv. As on date, CBBO-AAU is promoting 43 FPOs in 18 districts of Assam.
v. Out of the 43 FPOs, 38 are financially supported by NABARD as its implementing agency while remaining are supported by NCDC.
vi. 38 FPOs have been registered as Farmer Producer Companies (FPCs) under the Companies Act while five FPOs have been registered under the state cooperative act.

vii. All the FPCs/FPOs under CBBO-AAU have different focus commodities for business such as Red Rice, Joha Rice, Maize, Vegetables, Litchi, Mustard, Assam Lemon, Organic tea, Orthodox tea, Banana, Oil Palm, Pineapple, Black Rice etc.

viii. One FPC located at Sonitpur district (Agnigarh FPC Ltd.) has exported Litchi to the United Kingdom while another FPC located at Darrang district (Seujmukhi FPC Ltd.) has exported Dolichos bean to the UAE and Assam lemon to the United Kingdom.

ix. Through convergence with the state department of agriculture, more than 20 FPOs have acquired farm machineries under the Village Level Farm Machinery Bank scheme (central sector). Many of them have established Custom Hiring Centres.

x. Four FPOs have acquired combine harvester under the Chief Minister's Samagara Gramma Unnayan Scheme while two FPOs have acquired mini pick-up vans under the same scheme.

xi. One FPO of Lakhimpur district (Basundhara Agro Tech Co. Ltd.) have acquired honey processing equipments supported by the district administration. They are currently marketing honey under their own brand name.

xii. Two FPOs (Manikpur Joha Rice Producer Co. Ltd. Of Bongaigaon district and Besimari FPC Ltd. Of Darrang district) are producing chips grade potatoes under contract farming agreement with PEPSICO.

xiii. North West Jorhat FPC Ltd. of Jorhat district is implementing millet project under the millet mission with the financial support from NABARD and have marketed several value added millet products under the brand name 'Sri Bhumi'.

xiv. Nasiriba FPC Ltd. of Goalpara district is doing business in fresh banana and have marketed banana to other states like Bihar, West Bengal and UP.

xv. CBBO-AAU have been building the capacity of the nurtured FPOs and handholding them in establishing business linkages. Training for the Board of Directors (BODs), Chief Executive Officers (CEOs) and shareholder farmers have been organised for all the FPOs. Skill training and exposure visits have also been organised for most of the FPOs. Members of Agnigarh FPC Ltd. of Sonitpur district were even taken for exposure visit cum training at the National Research Centre for Litchi, Muzaffarpur, Bihar.

xvi. Nabajagaron Mahila FPC Ltd. of Jorhat district is a unique FPC promoted by CBBO-AAU as it happens to be an all-women FPO with more than 400 women members.

xvii. Different FPOs under CBBO-AAU has produced and marketed several value added products such as millet powder, mustard oil, honey, vermicompost, organic tea, red rice, black rice cake etc.

xviii. About four FPOs have already established retail fertilizer and pesticide business after acquiring necessary licenses.

xix. Two FPOs have started seed marketing business after acquiring seed marketing license.

c) Open and Distance Learning - A new initiative by Assam Agricultural University (AAU-ODL) for skill training

During the period of COVID-19 pandemic, most of the youth lost their jobs and returned to their villages in Assam. For many of them, agriculture and allied sector the only hope for the future to provide them a stable source of livelihood. However, to start with such a new means of earning, most of them need to first acquire the much-needed basic skills related to the vocation but they don't have access to suitable courses of short duration through which they can acquire such skills without visiting the formal corridor of institutionalized learning. Skill development in agriculture and allied sector is, therefore, ever-increasing and the basic requirement of education and expertise can be fulfilled by the e-courses if made accessible to the people who need them through open and distance learning programs. Assam Agricultural University initiated Open and Distance Learning (ODL) mode of education (AAU-ODL) with an overall aim of 'reaching the unreached'. It has started working with the objectives of providing holistic agricultural knowledge and skill amongst different categories of learners including farmers, entrepreneurs, rural women as well as unemployed youths, employees of Government/private sectors, field service providers, and school dropouts etc.

d) Promotion of double cropping in Assam through farmers' participatory approach

Rice fallow areas provide good scope for crop intensification as well as diversification and their suitable and efficient utilization can help in overcoming many social and economic problems including hindrance in employment generation, labour migration, low income and poverty. These areas can be effectively utilized for cultivation of suitable pulse crops *viz.,* lentil, lathyrus, field pea, rajmah, chick pea as well as oilseeds *viz.,* rapeseed and mustard, niger, linseed etc., during *rabi* seasons after *sali* rice. But, to utilize the residual soil moistures efficiently and effectively to maximize the system productivity in rice–fallows, long-duration rice varieties need to be replaced with medium duration varieties for early harvesting and timely sowing of succeeding crops. Farm mechanization is an important area as there is always scarcity of labour during the peak growing season and use of farm machineries not only reduces the cost of production but also helps in timeliness of different operations.

There is a huge potential to develop value added products from rice. The value added products having market potential are rice flour, rice flakes, rice bran oil, particle boards from straws, starch from broken rice etc. Further value addition can also be done in the form of preparation of rice confectionaries *viz.,* brown rice, puffed rice, flaked rice, ready-to- eat foods and extruded foods. Pulse production should be backed with quality post-harvest processing and storage system for attracting youths in this sector. Oilseed production can be another area which will create more employment in the entire chain from production to processing, branding and marketing.

Assam Agricultural University is trying relentlessly to promote double/triple cropping in the state and developed suitable technologies such as medium duration rice varieties and package of practices for different crops so that two or more crops can be cultivated in a year through efficient utilization of resources. In 2024-25, Assam Agricultural University initiated a program for promotion of Double Cropping in 1000 ha area across the state in 25 districts of Assam. The major objectives of the program are promotion of double cropping through efficient utilization of rice–fallow areas for increasing cropping intensity in the state; demonstration of production technologies of crops for up-scaling the area under double cropping; and increasing productivity and profitability of rice-based system for enhancing farmers' income.

e) Farmers' participatory technology showcasing and seed production program

Lack of quality seeds and planting materials is the major hindrance to increasing the production/ productivity of crops in Assam. In order to mitigate this, AAU launched an ambitious program 'Technology Showcasing' in 2010-11 through selected KVKs in farmers' participatory approach. Under the program, an area of 1100 ha was taken up by the selected KVKs for seed-production-cum-showcasing of high yielding crop varieties with scientific seed production technology. The main objectives of this program were :

- Producing quality certified seeds from foundation seeds in order to meet the shortage of certified seeds amongst the farmers.
- Utilizing the seed production process as a frontline demonstration in

order to influence and motivate farmers of the district so as to accelerate the adoption of scientific crop cultivation practices.

With this initiative, seed production of high yielding rice varieties was taken up in different KVKs and research station with farmers' participatory approach resulting in certified seed production as well as increasing the productivity levels.

f) Self-financed training program for agricultural input dealers to enhance technical proficiency

Agri-input dealers were playing a crucial role as primary sources of information for farming community, in addition to supply of agricultural inputs. However, majority of these agri-input dealers lack formal education on agriculture which hinders their ability to deliver comprehensive support to farmers. In order to improve their technical proficiency in agriculture and to facilitate them to serve the farmers better, Assam Agricultural University has introduced self-financed training program of 15 days' duration for input dealers with the help of KVKs across the state of Assam with the objectives of enhancing the technical skills and knowledge of input dealers so that, they can better serve farmers with their agricultural needs. By empowering input dealers, self-financed training program aims to enhance the reach of extension services and agricultural knowledge to a wider farming community, especially in rural areas of the state.

g) Community-led approach for enhancing climate resilience in adopted village

Indian Council of Agricultural Research (ICAR) on 2nd February 2011 initiated the project 'National Initiative on Climate Resilient Agriculture' (NICRA) with funding assistance from Ministry of Agriculture, Government of India along with Krishi Vigyan Kendra Dhubri under the authority of Assam Agricultural University at Udmari Part-IV and Part-V villages of Dhubri district of Assam. Rice is the most significant crop cultivated in Udmari village, grown primarily during the *kharif* season and also as an early summer (ahu) crop. However, flooding creates a major challenge-disrupting *Kharif* rice during its growth phase and also affecting summer rice during harvest. Submergence of crop for longer period during rainy season because of poor drainage results in crop loss of *kharif* rice up to 20 to 100 per cent depending on submergence duration. Likewise, early flooding during June or July often prevents farmers from transplanting rice seedlings on time, as poor drainage remains a major constraint. This delay leads to late transplanting which significantly reduce the yields. To address the issue of excess water accumulation in the crop fields, a targeted intervention was undertaken to renovate a drainage channel.

This drainage channel was previously dug out long back; over time, due to inadequate upkeep, the channel had become clogged with earth material leading to water stagnation or complete crop field submergence. Initially, some local people were against the construction of drainage channel as the channel would move through their own land. KVK Dhubri constituted a Village Climate Risk Management Committee (VCRMC) with the local people and commenced a move for renovating the drainage channel with due consultation with the local people. After several round of discussion, finally they agreed to the proposals and decided to extend full support to the officials. The supervision committee of KVK and VCRMC, NICRA project for construction oversaw the work for drainage channel renovation and applied the project by connecting the farmers' field to the river 'Gaurang' with a 1000 m long channel. The local people developed a rural road to the field by the side of the channel by utilizing dug out soil material. As an immediate outcome of the renovation work, no water submergence was there in the summer rice/early ahu rice crop field and crop was not damaged by water even though occurrence of heavy rain during April-May, 2013, and making it possible for the farmers to harvest their crop without any damage. Likewise, excess water during *kharif* season was quickly removed by this drainage channel and submergence was reduced which could save *kharif* crops.

The drainage channel aided the farmers from the villages under NICRA program along with some other adjacent villages which covered an area more than 500 ha. Moreover, an approach road was also developed to the crop field with dug out material. Some vital outcome of the channel is rapid and easy flood water drainage which permits farmers for timely growing of rice crop during *kharif* season and also reduces crop loss in *kharif* season due to remaining of flood water in crop field for short periods. The drainage channel benefitted a total area of 500 ha (both in kharif & summer season). The farmers are allowed to harvest summer/early ahu rice if flood occurs which augmented the average net return and benefit cost ratio from about Rs. 5,000 to Rs. 23,000 per hectare and from 1.3 to 1.8, respectively. The total area under rice cultivation has expanded during both the *kharif* and summer seasons, leading to an increase in cropping intensity to 150%. Additionally, the drainage channels, which remain filled with water during the rainy season, are now being effectively utilized as waterways for transporting harvested produce.

h) Linking producers to other support and services - Skill development, technology application and production

Assam is the abode of numerous highly nutritive indigenous fishes. However, due to the environmental incongruities, commercial exploitation of natural habitations and cultural malpractices, the availability of such fishes are decreasing. Hence, it is highly necessary to protect and increase these aquatic species by all potential and efficient breeding tools. Krishi Vigyan Kendra of Nalbari district of Assam is putting high priority in induce breeding and seed production of different indigenous fish. Since 2014, the programs have been working for diversification of cultivable fish species in aquaculture having high market demand along with rich nutritive value. One of the indigenous fish having high nutritive and market value is Magur (*Clarias batrachus*).

Magur is a catfish traditionally renowned for its healing value even. The price of this fish in the local markets during last few years has been ranging from Rs 400-1500 per kg. With some slight modification of the induce breeding technique of magur developed by ICAR-CIFA, Bhubaneswar, KVK Nalbari have started imparting hands on training to interested fish farmers of the district so that they become adept enough to start breeding of magur. Three farmers in the year 2014, initiated Magur fish breeding following acquirement of skill in the training. The keen interest of these farmers in seed production of Magur, encouraged KVK Nalbari for collaboration with ICAR-CIFA, Bhubaneswar for procurement of portable FRP magur hatcheries for them. The victory of those peer farmers in breeding and nurturing of magur fish caught attention and fascinated other fish farmers of the district in the noble operation.

In 2015, five successful magur breeders were again selected for FRP magur hatcheries and subsequently they were provided in the next year. The support given in terms of hands on training and FRP magur hatchery distribution worked wonder as these farmers could be able to breed magur and rear those seeds effectively. Out of capable beneficiaries (farmer friends) for the last two years, one could produce an average 70-85 thousand numbers of magur fingerling as ready for stocking in grow out pond. The calculated net average income per season is about Rs 2 lakh from each magur hatchery owner through seed production. The success of these farmers in induce breeding and seed rearing is establishing the magur culture an option of self-employment and entrepreneurship development for interested farmers of the district as well as in the state.

Nalbari district is now considered as a centre of attraction and place of availability of magur seed as farmers from different districts of Assam namely Jorhat, Bongaigaon, Sivasagar, Kamrup, Lakhimpur, Sonitpur, Darrang etc.

are obtaining seeds of magur from these farmers. The availability of seeds safeguards the magur fish farming in the state which eventually helps in fulfilling the demand for these nutritive fish. Another important impact is that the success of induce breeding reduces the pressure of wild catch of magur fish during April to July which is declared as "fishing ban season in the natural water bodies of the state of Assam." Thus, it indirectly upsurges in the natural stock of magur fish in natural water bodies like beels of Assam.

i. Use of ICT in agriculture

Agricultural advisories are one of the critical farm inputs for any farm activities, of which weather and climatic information plays an important role in agriculture. If provided in advance, it will help farmer in decision making and efficient mobilization of farm resources. KVKs under AAU have been delivering information to the farming community through Kisan Mobile Advisory Services harnessing the ICT based mobile services with the objective of providing need based information related to agriculture and allied sectors timely.

The Assam Agricultural University (AAU), Jorhat, has been offering the service of Agro-met advisories since 1994 with the introduction of the AAS scheme at its headquarters in Jorhat, rendering its service in the districts of Upper Brahmaputra Valley Zone (UBVZ) of Assam. Later, it was extended to all six Agrolimatic Zones of Assam under the IAAS and GKMS scheme. The locations specified for rendering these services in Assam are named as AMFUs and DAMUs situated in the Zonal Research Station (ZRS) and Krishi Vigyan Kendra's (KVK's) of Assam Agricultural University. The designated location of AAU currently involved in preparation and dissemination of weather based agro-met advisory services are listed below:

S. No.	Agro Climatic Zones	AMFU's
1.	Upper Brahmaputra Valley Zone (UBVZ)	College of Agriculture, AAU, Jorhat
2.	North Bank Plain Zone (NBPZ)	BN College of Agriculture, AAU, Biswanath
3.	Central Brahmaputra Valley Zone (CBVZ)	AAU-ZRS, Shillongani
4.	Hill Zone (HZ)	AAU-ZRS, Diphu
5.	Lower Brahmaputra Valley Zone (LBVZ)	AAU-ZRS, Gossaigaon
6.	Barak valley Zone (BVZ)	AAU-ZRS, Karimganj

A regular user of this service get a brief picture of the past week weather condition along with the forecasted five days' weather and crop/livestock/poultry/fisheries information (like type, stage, pest-disease or any other stress status, etc. Such information is value added with some specific climate-smart crop advisories which will either enhance the adaptive capacity of the farmers to counter adverse weather conditions or will help to optimize the use of available resources.

Suggestions

i. Innovative and ground-breaking extension approach must be worked on to make agriculture more profitable and sustainable imparting livelihood security to the farmers. In this new era, transformation should be aligned with technology driven extension system to market driven system.

ii. Sometimes, technologies are less effective or not adopted by the farmers because of the complex nature or as the required inputs are not readily accessible to the farmers. So, coordination of researchers, extension personnel and farmers are required to identify researchable issues for generating need-based technologies as well as dissemination of feasible technology

iii. In the pluralistic extension system which prevails in the country with public sector, private sector and third sector, all the sectors should work in coordination as the objectives are same.

Conclusion

Strong extension system plays an important role to meet the present day challenges in agriculture. Agricultural extension is playing a wider role by developing human and social capital, enhancing skills and knowledge for production and processing, facilitating access to markets and trade, organizing farmers and producer groups, and working with farmers for sustainable agricultural development. It needs transformation to provide a diverse set of services that support agricultural livelihoods, offering relevant technologies that are integrated with appropriate services.

References*

1. Annual Report. 2016-17. Government of India, Ministry of Agriculture & Farmers Welfare (Department of Agriculture, Cooperation & Farmers Welfare) Directorate of Pulses Development, Vindhyachal Bhavan, DPD/Pub/TR/19/2016- 17, pp 46–7.
2. Mahapatra. 2020. India's agrarian distress: Is farming a dying occupation Down to Earth. Thursday 24 September, 2020. https://www.downtoearth. org.in/ news/agriculture/ india-s-agrarian-distress-is-farming-a-dying-occupation -73527
3. Praharaj, C.S., Kumar, R. and Blaise, D. 2016. Intercropping: An approach for area expansion of pulses. *Indian Journal of Agronomy* **61**: S113-S121.
4. Statistical Hand Book of Assam, 2023, Directorate of Economics and Statistics, Govt. of Assam.

* The content of this chapter was prepared in consultation with above references

20

Emerging Lessons from Farmers Producers Organisations – A Case Study of FPOs of District Budgam, UT of J&K

Bhinish Shakeel[1], Bilal A. Lone[2], Vaseem Yousuf[3], Rafiya Munshi[4], Mir Nadeem Hassan[5] and Ambreen Nabi[6]

Senior Scientists, Shere Kashmir University of Agricultural Sciences and Technology-Kashmir

Email: bhinishshakeel@skuastkashmir.ac.in

Abstract

The rhetoric on Farmers Producers Organisation (FPOs) as an assemblage of farming community has been gaining a lot of popularity off-late. The FPOs are essentially meant to strengthen small and marginal farmers with an end-to-end support and service including provide small farmers with end-to-end support and services, including technical assistance, marketing, processing, and cultivation inputs. The core philosophy is to raise the standard of these farmers by facilitating their income. FPOs which are formed as Farmer Producer Companies (FPCs) allow members to access financial and other input services. To compete in market, FPOs must be competent with other companies and competitors in the market, and they have a tremendous potential to capture future food retails not only in India but throughout the world. Keeping this in view an attempt has been made in this book chapter to highlight various constraints related to the growth, performance, and challenges of FPOs along with strategies to make them more effective in the present context.

Keywords: *Challenges, Farmers Producers Organisations, Jammu and Kashmir*

Introduction

In the paradigm of development, the Government of India generated a Central Sector Scheme for the people associated with agriculture – a sector contributing to 18.2 per cent of the nation's GDP, titled "Formation and Promotion of 10,000 Farmer Produce Organizations (FPOs)" with major focus on upheaval of farming community. With well-defined strategy and committed resources the Scheme is set-out to formulate 10,000 new FPOs in the country with budgetary provision of Rs 6865 crore.

At the core of formation of each FPO, that are to be developed in produce clusters, wherein agricultural and horticultural produces are grown / cultivated for leveraging economies of scale and improving market access for members. "One District-One Product" cluster to promote specialization and better processing, marketing, branding & export. Under this Central Sector Scheme with funding from Government of India, Formation & Promotion of FPOs are to be done through the Implementing Agencies (IAs). Presently 09 Implementing Agencies (IAs) have been finalized for formation and promotion of FPOs viz. Small Farmers Agri-Business Consortium (SFAC), National Cooperative Development Corporation (NCDC), National Bank for Agriculture and Rural Development (NABARD), National Agricultural Cooperative Marketing Federation of India (NAFED), North Eastern Regional Agricultural Marketing Corporation Limited (NERAMAC), Tamil Nadu-Small Farmers Agri-Business Consortium (TN-SFAC), Small Farmers Agri-Business Consortium Haryana (SFACH), Watershed Development Department (WDD)- Karnataka, and Foundation for Development of Rural Value Chains (FDRVC)- Ministry of Rural Development (MoRD).

The Implementing Agencies (IAs) engage Cluster Based Business Organizations (CBBOs) to aggregate, register & provide professional handholding support to each FPO for a period of 5 years. CBBOs have been empanelled & engaged by IAs. CBBOs operate as an end-to-end knowledge for all issues related to FPO promotion. During 2020-21, a total of 2200 FPO produce clusters were allocated for formation of FPOs, which also include specialized FPO produce clusters such as 100 FPOs for Organic, 100 FPOs for Oil seeds, etc. Of these, 369 FPOs are targeted for 115 aspirational districts in the country. Additionally, NAFED would form the specialized FPOs which should necessarily be forwardly linked to the market, agri-value chain, etc. NAFED will provide market and value chain linkages to the FPOs formed by other Implementing Agencies. NAFED has formed & registered 05 Honey FPOs during current year in Uttar Pradesh, Madhya Pradesh, Rajasthan, Bihar and West Bengal.

FPOs are earmarked to provide financial assistance upto Rs 18.00 lakh per FPO for a period of 03 years. In addition to this, provision has been made for matching equity grant upto Rs. 2,000 per farmer member of FPO with a limit of Rs. 15.00 lakh per FPO and a credit guarantee facility upto Rs. 2 crore of project loan per FPO from eligible lending institution to ensure institutional credit accessibility to FPOs.

At district level, a District Level Monitoring Committee (D-MC) is constituted under the Chairmanship of District Collector/ CEO/ ZillaParishad with representatives of different related departments and experts for overall coordination & monitoring the implementation of scheme in the district including the suggestion for potential produce cluster & development

At National level, National Project Management Agency (NPMA) as a professional organization has been engaged for providing overall project guidance, coordination, compilation of information relating to FPOs, maintenance of MIS and monitoring purpose.

There are well defined training structures in the scheme and the institutions like Bankers Institute of Rural Development (BIRD), Lucknow and Laxman Rao Inamdar National Academy for Co-operative Research and Development (LINAC), Gurugram have been chosen as the lead training institutes for capacity development & trainings of FPOs. Training & skill development modules have been developed to further strengthen the FPOs also through CBBOs.

Formation & promotion of FPOs is the best foot forward to convert Krishi into Atmanirbhar Krishi. This will enhance cost effective production and productivity and higher net incomes to the member of the FPO. Also improve rural economy and create job opportunities for rural youths in villages itself.

FPOs Formulated Under NCDC at the Krishi Vigyan Kendra – Budgam

In the year 2022, KVK-Budgam was empanelled as CBBO by the NCDC to formulate two vegetables and mushroom based FPOs in the District. After subsequent community mobilisation through awareness programs in the District, the Kendra succeeded in assembling farmers in Block Budgam and Bagat-e-Kanipora (B.K. Pora) as Sheikh ul Alam and Umeed Ki Kiran respectively in Feb. 2022.

Indicator	**Umeed Ki Kiran Farmer Producer Cooperative Ltd**	**Sheikh Ul Alam Farmer Producer Cooperative Ltd**
No. of Farmer Members	114	136
Equity Grant recieved on 1st Tranche	2.05 Lakh (Received)	1.66 Lakh (Received)
Management Cost Availed/ Applied for	1st Tranche: 3.66 Lakh (Received)	1st tranche: 3.48 Lakh (Received) 2nd trenche:2.95 lakhs(Applied)
Registered on e-NAM portal	Yes	Yes
Office Bearer Appointed	Yes	Yes
Status	Functional	Functional

Technologies Addressed through Extension Outreach

As part of handholding to these FPOs, KVK Budgam has been rendering Extension Outreach Services to these Farmers through community mobilisation for FPO membership, short-duration Training and Awareness Programmes on Protected Cultivation of Vegetables, Nursery Raising of Vegetable Crops as Profitable Venture, Scientific Cultivation of Exotic Vegetables, Pollination Management in Cucurbitaceous Crops.

Similarly, several Schools on Fruit and Vegetable Processing, Dhingri Mushroom Cultivation, Hybrid Seed Production in Vegetable Crops, Packaging, Labelling and Licensing in Processed Fruits and Vegetables were organized. All these aspects are dealt with practical demonstrations and exposure visits in line with Principles of Extension for effective learning. In prevailing competitive times

Apart from these, the farmers have been hand-held for packaging and labelling in processed fruits and vegetables to sharpen their repertoire of generic skills required to deal with exhibitions at regional and national level.

Fig 1. Community Mobilisation

Fig 2. General Body Meeting

Fig 3. Trainings and Schools

Fig 4. Exhibition at Annual Seed Mela SKUAST-K

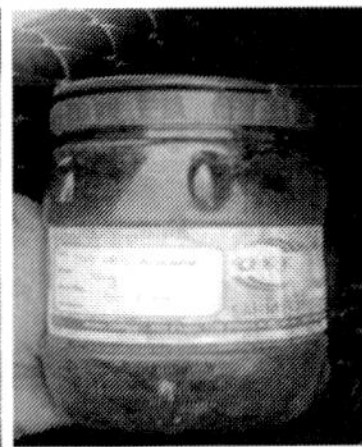

Fig 5. Products generated by FPO

Fig 6. Introduced Pared and Peeled vegetables for home delivery (Order based) in vicinity

Challenges Faced by FPOs at Budgam

Despite growing popularity of FPOs for strengthening farmer's groups at the grass-roots some bottlenecks have been identified as follows:

S. No.	Issues	Possible Solution
1.	Tough eligibility criteria for Seed/ Fertilizer License.	The FPOs need to avail a barrage of licenses to be a commercially viable group. The Board of Directors and the member FPO may or may not be science graduates. There is a need for relaxation in eligibility criteria to obtain Seed/Fertilizer/Pesticide license to facilitate preliminary formalities for execution of FPO. Also, there is no waiver in registration fee for these groups.
2.	Tedious Process of Registration for Licenses	For execution of FPO, there are certain licenses that are laid down under the guidelines to obtain TIN, PAN, GST, Seed/ Fertiliser/Pesticide, E-NAM, ONDC, fssai, APEDA, Mandi license. It is a cumbersome and tedious process.
3.	Conflict of interest among Board of Directors	Rivalry among members to achieve key positions in the organization and challenging each other for key positions in the group have been observed. Often the members are reluctant to give valid input in operations to CEO and other members with the view that those who are monetary remunerated must be ones who must be solely accountable for operations.
4.	Non-inclusion of Local Elected Leaders	The Board members are farmer's representative from the region of FPO. CEO and Accountant appointed for business operations may or may not be from the same geographical area. Non-inclusion of Local Elected Leader in the FPO which could act as a facilitator in case of conflict in the group.
5.	Lack of Exposure and professional management and among FPO Members	Most members of FPO are illiterate or semi-literate. Education, not to be viewed only as functional literacy, broadens the horizons of an individual. In absence of exposure to Co-operatives, Best-Practice FPOs in other areas that can be emulated, the FPOs do not have a heads-up as to how to lead, develop a stronghold, develop a niche with some innovative service, to sustain as a farmer's collective. Lack of cohesion between FPO Members has also been observed. More often they are not privy on various activities of FPO.
6.	Absence of Skill Sets in Board of Directors/Members of FPO	As already stated, the farmers are illiterate and are not equipped with skill sets, particularly Soft Skill Sets that could groom them in handling inter-personal dynamics with members. Moreso, the members are inept to deal with technology, particularly Information Communication Technology. (ICT)

The same constraints and issues have been reported in the Maharashtra, Osmanabad District in 2019 where members reported non-inclusion of local leaders in FPCs, lack of coordination for different group activities, lack of support from the government department after the establishment of FPCs, political affiliation of members, companies have limited access to banks, inadequate profit to individual members and village-level workers not providing enough information about all schemes related to FPCs.

A study conducted by Navaneetham et al (2019) on Analysis of constraints for performance improvement of FPCs in Tamil Nadu revealed that capturing the market for selling the produce was the biggest constraint and farmers were unable to raise funds from farmers.

Similarly, in a Strategy Paper for promotion of 10,000 Farmer Producer Organisations (FPOs), by SFAC (2019) some challenges faced in the promotion of FPOs are: difficulty and delay in the mobilisation of farmers; limited organisational and management capacity of FPOs; need for incubation and handholding support to FPOs; membership base of an FPO; policy level challenges; limited capability to autonomously invest in primary / secondary processing; storage and custom hiring facilities; and inability of FPOs to access institutional credit sans collateral.

Poor professional management, shortage of working capital, inability to access loans from financial institutions, unawareness of producer-members, insufficient directions and vision from the Board of Directors and poor infrastructure facilities were major hurdles for better performance of Farmer Producer Organizations as revealed in a study on socio-economic impacts on members of FPO in the plains region of Chhattisgarh reported by Prishila Kujur *et al* (2019).

Verma *et al* (2021) conducted a study on Constraints perceived by the members and non-members towards the functioning of FPO-AKPCL in Kannauj District of Uttar Pradesh. The results revealed that inadequate storage facilities, shortage of transportation facilities, lack of grading and packaging skills, rivalry among members to achieve key positions in the organization, and challenging each other for key positions in the group were the significant constraints faced by the member farmers.

According to Chauhan et al (2021), the constraints associated with the functioning of FPOs were undeveloped storage and processing facilities, lack of computer knowledge due to which they are unable to derive benefits of the available ICT tools, lack of awareness about packaging, lack of labour available during harvesting, lack of sufficient finance, lack of skilled labour in harvesting, processing, fluctuation of price every year, lack of proper market information, involvement of middlemen and lack of proper infrastructure. The study was conducted in Cooch Bihar district of West Bengal by collecting primary data from 100 FPO members.

Conclusion

Despite major thrust of government on formation of FPOs by Government as a launch-pad for small and marginal farmer's socio-economic development, the FPOs are far from achieving this goal in view of the challenges mentioned in this narrative. There is a need for intensive rigorous capacity building in terms of strengthening these groups in terms of human resource (role clarity, inter-personal dynamics, ownership, collective responsibility etc). Trainings at KVKs and other centres needs to be supplemented strongly with exposure visits to model FPOs for cross-learning as would give some food for thought to emulate in their own region. All the support to FPOs must be sustained rigorously till it achieves it logical conclusion of an economically viable farmer's collective.

References

1. Chauhan. K., Adhikary, A., and Pradhan, K. (2021). Identification of Constraints Associated with Farmers' Producer Organisations (FPOs). International Journal of Current Microbiology and Applied Sciences, 10(01): 1859-1864.
2. Chopade., S.L., Kapse, P. S., and Dhulgand V. G. (2019). Constraints faced by the members of Farmer Producer Company, International Journal of Current Microbiology and Applied Sciences, 8(8): 2358-2361
3. Navaneetham, B., Mahendran, K., and Sivakumar, S.D. (2019). Analysis of constraints for performance improvement of FPCs in Tamil Nadu. International Journal of Farm Sciences, 9(2): 12-18
4. Prishila Kujur., Gauraha, A. K., and Netam, O. K. (2019). The socio-economic impact of farmer producer organizations in Chhattisgarh plains. Journal of Entomology and Zoology Studies, 7(6): 1104-1106
5. SFAC (2019). Strategy Paper for promotion of 10,000 Farmer Producer Organisations (FPOs). New Delhi SFAC.
6. Verma, A.K., Singh, V. K., Asha, K., Dubey, S. K., and Verma, A. P. (2021). Constraints Perceived by the Members and Non-members towards Functioning of FPO-AKPCL in Kannauj District of Uttar Pradesh. Economic Affairs, 66(2): 335-341.

21

Re-affirming Faith in Agriculture Systems - Tribal Youth Embraces Integrated Farming System as Livelihood

Bhinish Shakeel[1], Bilal A Lone[2], Mir Nadeem Hassan[3], Vaseem Yousuf[4], Ambreen Nabi[5], Shabeer A. Ganie[6], Sabia Akhter[7], Shazia Ramzan[8], Iram Farooq[9], Khurshid Zargar[10], Shabeena Majid[11] and Asima Amin[12]

Senior Scientists, Shere Kashmir University of Agricultural Sciences and Technology-Kashmir

Email: bhinishshakeel@skuastkashmir.ac.in

Abstract

Mr. Mudasir Ahmad Dar, who has carved a niche for himself in the field of agriculture through adoption of Integrated Farming System (IFS) in the Tribal region of District Budgam. He has exemplified a working model of IFS in a short span of time dovetailing benefits of various government assistance in order to maximize yield and productivity per unit area through intensification and diversification of crops with integration of allied agri-based enterprises. He has been hand-held by Krishi Vigyan Kendra - Budgam along with residents of Shonglipora, Khag, a far-off Tribal region in district Budgam. Krishi Vigyan Kendras Budgam, operating as District resource centre for dissemination of scientific technology on agriculture and allied sciences, since its inception in 2013, has been constantly endeavouring to train and support rural entrepreneurs to upscale their skill sets.

Keywords: *Allied agri-based enterprises, Integrated Farming System, Krishi Vigyan Kendra*

Introduction

Rural entrepreneurship is not only important as a means of generating employment opportunities in the rural areas with low capital cost and raising the real income of the people, but also its contribution to the development of agriculture and urban industries Building capacity of Rural Agri-prenuers are a promising solution for unemployment (Renuka, 2020), migration, economic disparity, reduce poverty, development of rural areas and backward regions. Most of the rural populace is associated with agriculture and allied activities as pedigree occupation. However, to be commercially viable, these units need to sharpen their skill sets to thrive as a successful venture.

India is home to about 700 tribal groups with a population of 104 million, as per 2011 census. This indigenous people constitute the second largest tribal population in the world after Africa. Tribal population of Jammu & Kashmir is among the nascent Tribal Groups joining the main stream of planned development, to which they have brought a distinct and colourful cultural variety. Their economy is closely linked with the forests and they are living a substandard life because of their primitive mode of livelihood. Majority of them are placed below the poverty line, possessing meagre assets and are exclusively dependent on wages, forest produce and farming, that too in a traditional way which leads to non remunerative returns. The peculiar aspect of tribals of our state is their scattered population who inhabit the difficult and remote geographic terrains which poses a severe threat to their speedy development. In J&K state the following communities have been declared as scheduled tribes 1. Balti, 2. Beda, 3. Bot, Bota, 4. Brokpa, Drokpa, Dard, Shin, 5. Changpa, Garran, 6. Mon, 8. Purigpa, 9. Gujjar, 10. Bakerwal, 11. Gaddi and 12. Sippi.

Factors Enabling Favourable Agri-Development in Tribal Areas (SWOT Analysis)

Strengths

1. Traditional knowledge: Tribal farmers in India possess traditional knowledge and skills related to farming and agriculture, which have been passed down from generation to generation.
2. Local varieties of crops: Tribal farmers cultivate a variety of crops that are well-suited to the local agro-climatic conditions and have high nutritional value.
3. Sustainable farming practices: Tribal farmers have a deep understanding of the local ecosystem and practice sustainable farming methods, which help to conserve natural resources and maintain the ecological balance.
4. Strong community ties: Tribal communities in India have a strong sense of community and social cohesion, which enables them to work together towards a common goal.

Weaknesses

1. Limited access to resources: Tribal farmers in India often have limited access to resources such as land, water, credit, and infrastructure.
2. Many tribal areas in India lack basic infrastructure such as roads, electricity, and storage facilities, which makes it difficult to transport and store agricultural produce.
3. Low levels of education: Tribal farmers often have low levels of education and may lack the skills and knowledge required to take advantage of new technologies and market opportunities.
4. Vulnerability to climate change: Tribal farmers are often located in remote areas and are particularly vulnerable to the impacts of climate change, such as droughts, floods, and crop failures.

Opportunities

1. Government support: The Indian government has launched several schemes and programs to support tribal farming, such as the Integrated Tribal Development Program and the Tribal Sub-Plan.
2. Growing demand for organic produce: There is a growing demand for organic produce in India and abroad, which presents an opportunity for tribal farmers to increase their income.
3. Value addition: Tribal farmers can add value to their agricultural produce by processing and packaging it, which can lead to higher profits and better market access.
4. Tourism: Tribal areas in India are known for their unique culture and traditions, which can attract tourists and provide an additional source of income for tribal farmers.

Threats

1. Land acquisition: Tribal farmers in India are often at risk of losing their land to large-scale development projects such as mining, dams, and highways.
2. Market competition: Tribal farmers face competition from larger farmers and agribusinesses, which have greater access to resources and market information.
3. Changing dietary habits: Changing dietary habits and preferences among consumers may lead to a decline in demand for traditional crops grown by tribal farmers.

4. Climate change: Climate change poses a significant threat to tribal farming in India, as it can lead to crop failures, loss of livestock, and damage to infrastructure.

Extension Outreach and Initiatives

Khag is a tehsil (since 2008) in Beerwah Sub-district of Budgam which is approximately 20 km from Budgam and 35 km from District Srinagar. It is connected to Mazhama Railway Station (19km) with closest airport as Srinagar International Airport (55 km). Resident of village, Drang, the entrepreneur initiated the experiment a year ago with vegetable production under protected conditions (polyhouse), production of field crops and sheep rearing. As part of initiatives under the University outreach for progressive farmers led through the Directorate of Extension, a high value vegetable production was initiated under a high-tech polyhouse following a proper package-of-practice including Integrated Nutrient Management (INM), Integrated Disease Management (IDM) and Integrated Pest Management (IPM) etc as instructed by Krishi Vigyan Kendra Budgam. Adherence to Package of Practice laid down by SKUAST-K and availing Government Schemes of Assistance through Line Departments. IFS model of the farmer combines various compatible enterprises such as crops (field crops, horticultural crops). Resource utilization of resources efficiently is critical for sustainable development. It is only through judicious use of resources and inter-linking the enterprises, a farmer can ensure productivity, profitability, livelihood creation, conservation of natural resources and maintenance of sustainable ecosystem

Technologies Addressed

Integrated farming system is a sustainable agricultural system that integrates livestock, crop production, fish, poultry, tree crops, plantation crops and other systems that benefit each other. It is based on the concept that 'there is no waste' and 'waste is only a misplaced resource' which means waste from one component becomes an input for another part of the system. IFS approach is considered to be the most powerful tool for enhancing profitability of farming systems especially for small and marginal farmers to make them bountiful.

Mudasir Ahmad has aptly adopted the IFS and in this journey of the IFS, favorable collaborations between Line Departments, KVK and FVSc and AH, SKUAST-K that have yielded fruitful results. Capacity Building Programme on Wool and Pelt Processing in collaboration with Division of LPT, FVSc and AH, SKUAST-K, Capacity Building Programme Value Addition of Milk, in collaboration of LPM, FVSc and AH, SKUAST-K and Skill Training for Rural Youth on Protected Cultivation of Vegetable Crops have all aided this process of transformation of a Profitable IFS enterprise in a Tribal Area.

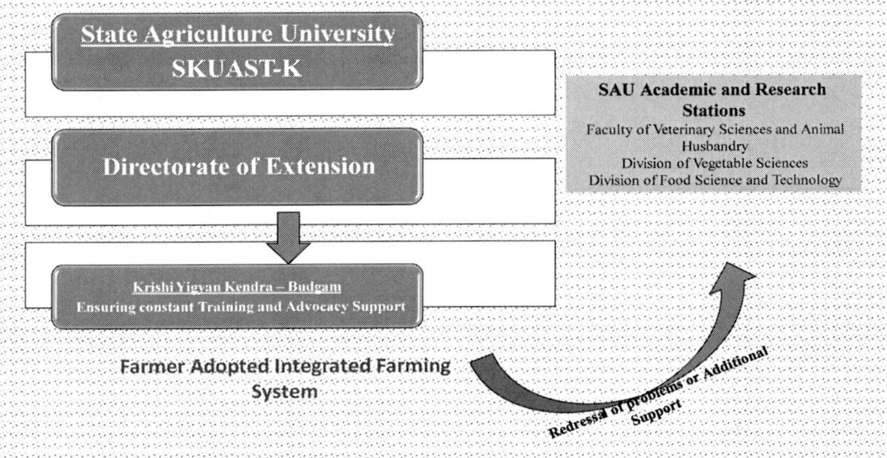

Fig 1. Model of operations and extension support

Impact of the Venture

Quantitative Outcomes

S. No.	Crop	Area (in Kanal) /No	Production (in quintal)	Gross Income (Rs.)	Net income (Rs.)	B:C Ratio
Field Crops						
01	Oats	2.0	2.0	12400	9400	3.13
02	Rice	13	49	122500	76500	1.66
03	Rajma	3.0	2.0	30000	24000	4.00
Horticulture Crops						
04	Onion	0.5	1.0	2800	2000	2.50
05	Garlic	1	5.0	35000	24000	2.18
06	Apple	12	360	476000	300000	2.27
07	Walnut	45 plants	7.5	150000	120000	4.00
08	Vegetable cultivationunder protected conditions(tomato, Cucumber, cherry & Capsicum	1	10	100000	700000	2.50
Livestock						
09	Cows	02	80000 ltrs	320000	230000	2.40
10	Sheep	20	@10000/sheep	200000	145000	2.63

The Krishi Vigyan Kendra – Budgam, further extended support to mobilise dairy farmers of the region for assemblage into a Dairy Cooperative for facilitate milk sales. Dairy farmers of the region (Khag, Nasrpura, Trapei, Shonglipora, Bathipora, Sitaharan, Drang) were mobilised after successive mobilisation drives in the community into a Farmers Producer Organisation (FPO), The Khag Kissan Farmer Producer organization cooperative Ltd. Drung Khag sponsored by NABARD. Mudasir Ahmad Dar is constantly endeavouring for mobilizing farmers into the FPO, business planning, market linkages and other allied activities of the FPO as its Chief Execution Officer. The FPO has already generated assets like Automatic Milk Collection Unit (AMCU), Wool Shearing Machines and Wool Weighing Scales availing Government Assistance under Holistic Agriculture Development Plan that rolled out in 2022 as a massive UT Government Initiative for upheaval of agriculture and allied sector in Jammu and Kashmir. A vehicle worth 12 lakhs has also been purchased for milk supply has also been purchased for milk distribution.

Output: Asset Generation

Details	Assistance	Status
Bulk Milk Cooler	Holistic Agriculture Development Plan	06 No.
Automated Milk Collection Units	-do-	04 No.
Wool Weighing Scale	-do-	5 Quintal (02 No.) 2 Quintal (01 No.)
Wool Shearing Machine	-do-	03 No.
Fat Analyser	-do-	02 No.
Van	-	01 No

Fig 2. KVK Budgam Team mobilising Dairy Farmers

Fig 3. Activities under Dairy, Sheep Rearing and Vegetable cultivation

Fig 4. Activities under Vegetable Production, Dairy FPO and Wool Shearing

Impact of the Endeavor

- Building faith in agriculture as a promising livelihood option in times when most agricultural land is being rampantly diverted to non-agricultural uses
- Leading by example and becoming leader of sorts, encouraging local youth to emulate the experience
- Fostering a collaboration between Line Departments, NGOs, KVK and other resource centres of SKUAST-K
- Improving the standard of living through maximizing the total net returns and provide more employment, minimizing the risk and uncertainties
- Skill development through various trainings programs undertaken at KVK Budgam SKUAST-K

Challenges Faced

Marketability: The Division of Vegetable Science facilitated marketability of the farmers produce by providing an authorised market for sale of potato.

Under Production: Vegetable production was enhanced through awareness cum training programs by KVK Budgam, SKUAST-K and scientific/technical advice from scientists of KVK Budgam has led to optimum production of vegetable crops and seedling for sale thereof.

Challenges faced in milk production are highly perishable commodity with an unsure market. They are currently negotiating with the leading local Dairy Corporate, Khyber Agro for a sustained rate (Rs 42 per litre) instead of Rs 39 per litre.

Conclusion

Mr. Mudasir Ahmad Dar is the emerging face of an educated agri-prenuer. He has marked his presence in the field by taking the lead and consolidating activities as a commercial unit. He has a mention worthy annual turnover of Rs. in a short span of time. Having tasted success through sheer hard work, Mr. Mudasir inspires to take his venture to newer heights. He now intends to explore opportunities in value addition of vegetables and preservation of milk through canning/bottling, deep freezing of mutton etc. His journey is a testament that motivation to embrace agri-based livelihood seeking Government schemes and State Agricultural Universities support can transform rural economies into sustainable agri-economies.

Reference

1. Renuka, E., 2020. Role and Importance of Entrepreneurship Development in Rural Area. *The International journal of analytical and experimental modal analysis*, 8(2): 110-114.

22

Innovative Extension Approaches of KVAFSU, Bidar

K.C. Veeranna[1], L. Manjunatha[2], Shivakumar K. Radder[3], V. Jagadeeswary[4], Prakashkumar Rathod[5] and Channappagouda Biradar[6]

[1]Vice-Chancellor, Karnataka Veterinary, Animal and Fisheries Sciences University (KVAFSU) Bidar, [2,4]Professor, Veterinary College, Bengaluru, [3]Professor, Veterinary College, Gadag, [5]Associate Professor, Directorate of Extension, KVAFSU, Bidar, [6]Associate Professor, Veterinary College, Bidar, Karnataka, India

Email: prakashkumarkr@gmail.com

Abstract

Karnataka Veterinary, Animal and Fisheries Sciences University, Bidar, was set up with the mission to strive hard and provide leadership in teaching, research and extension education services related to Veterinary and Animal sciences, Dairy sciences, Fisheries and other allied sciences. Various extension educational activities are undertaken by the university at various levels, viz., Extension Education, Extension Research, Extension Work and Extension Service. Collaborative activities are being planned and implemented in association with various stakeholders like Farmers, Farmers' organizations, Department of Animal Husbandry and Veterinary Services, Karnataka Milk Federation etc. In course of time, along with conducting routine extension activities like trainings, demonstrations, animal health camps, field days, field trips, campaigns, livestock shows etc., KVAFSU has come out with some innovative approaches in extension education and these initiatives are categorized under "Innovative Extension Approaches through university/college collaborations and sponsored partnerships". Important highlights of all these are – Effective use of Performing arts as extension tools and Participatory Research and Extension approaches for sustainable livestock development.

Keywords: *Livestock Development, Performing arts, Participatory Research and Extension approaches*

Introduction

Karnataka Veterinary, Animal and Fisheries Sciences University (KVAFSU), Bidar started in the year 2005 exclusively for the development of education and learning; conduct of research and extension education and transfer of rural oriented technologies in the areas of Veterinary, Animal, Dairy and Fisheries Sciences in Karnataka. The logo of the University is inscribed with the slogan "Farmers Friendly and Rural Oriented", which emphasizes the importance given to betterment of farming community.

The mission of Karnataka Veterinary, Animal and Fisheries Sciences University (KVAFSU), Bidar, is to strive hard and provide leadership in teaching, research and extension education services related to Veterinary and Animal sciences, Dairy sciences, Fisheries and other allied sciences. The University endeavor is to keep pace with new frontiers of science and contemporary developments to be socially and technically relevant. In this context, KVAFSU is strongly committed in absorbing newer paradigms and using them to develop excellent human resource, innovative technologies and their dissemination so as to serve the livestock and fishery farming community of the state and the country.

Various extension educational activities are undertaken by the university at various levels, *viz.*, Extension Education, Extension Research, Extension Work and Extension Service. Collaborative activities are being planned and implemented in association with various stakeholders like Farmers, Farmers' organizations, Department of Animal Husbandry and Veterinary Services, Karnataka Milk Federation, etc. Besides conducting routine extension activities like trainings, demonstrations, animal health camps, field trips, field tours, campaigns, livestock shows etc., KVAFSU has come out with some innovative approaches in extension education. Which are presented in this book chapter. Extension activities are undertaken mainly under two categories, i.e. university / college activities and collaborative / sponsored activities.

I. Innovative Extension Approaches Under University / College Activities Organization of Radio Series for Farming Community

Radio remains an effective medium for disseminating veterinary knowledge, especially in rural areas. KVAFSU collaborates with radio stations to

- Broadcast expert talks on Veterinary and Animal Husbandry topics. More than 150 expert talks from Veterinary college Hassan, entitled as Gokulavani series, Kukkutavani series, Hainnu Honnu series I, II, II are broadcasted through AIR Hassan. Presently, Hainnu Honnu series IV is being executed with another 50 series of expert talks and skits as well.

- Disseminating scientific information amalgamated with entertainment in a skit format to farmers through AIR. More than eight skits have been broadcasted by students of Veterinary College, Hassan through AIR Hassan, and the same has been linked to FARMERS CORNER website of KVAFSU (https://www.kvafsu.edu.in/farmars_corner.html) for entire public usage of information.
- The expert talks are compiled in the form of a Book and released for public usage.
- Hosted interactive Q&A sessions with farmers for problem-solving.
- Share success stories of progressive farmers to motivate others.
- AIR advisory programs during disease outbreaks and seasonal challenges.
- AIR quiz sessions after each series, of expert talks and acknowledging the prize winners in college farmers programs.
- Promotes government schemes related to livestock development.

Fig. 1. Hassan Veterinary College students Radio Skit on "ASHAKIRANA - Women Empowerment through Animal Husbandry" in AIR Hassan"

Art Forms as Extension Education Tools in Livestock Development

Various art forms are very popular across the country. Besides entertainment, they have been a source of knowledge, heritage and inspiration especially to rural folks. Being part of customs and folklores of farmers, they have a lot of potential to further the cause of extension education in the field of Animal Husbandry and Veterinary Sciences. Keeping this in mind, faculty of the university along with students have tried some art forms as tools for dissemination of scientific information to livestock farmers. Two important tools *viz.*, songs and skits have been tried in this direction and some findings are detailed below :

Use of Songs / Popular Music Tunes for Dissemination of Animal Husbandry Information

Traditional media such as music, folk dance, etc., are being successfully used in developmental communication. In this context, a study was undertaken to utilize popular music tunes for dissemination of scientific animal husbandry management practices.

Five important animal husbandry management practices were chosen (using relevancy rating) and subsequently, appropriate popular music tunes in Kannada language that would sync with themes were selected (using relevancy rating by expert panel with both technical and musical background). Further, lyrics to match with the selected tunes were written, evaluated and finalized in consultation with the experts. Professional technical expertise was utilized for production of the songs. Thus, songs developed were tested for effectiveness by playing them to farmer participants in various training programs. A pre-tested structured schedule was used to assess the knowledge level of the farmers on the selected topics. The knowledge level of the sample respondents on the selected topics pre and post listening were collected and analysed using appropriate statistical tools.

The five important animal husbandry themes selected were: Importance of fodder production; Clean milk production; Vaccination; Deworming; Control of ectoparasites; Foot and Mouth Disease and Rabies disease. The songs developed on these were tested for effectiveness. A significant increase in the knowledge level was found among them after listening to the songs [fodder production: $t = 25.6$, $p < .001$; Clean milk production: $t = 23.3$, $p < .001$; Vaccination, Deworming and Control of ectoparasites: $t = 32.9$, $p < .001$; Foot and Mouth Disease: $t = 24.7$, $p < .001$ and Rabies disease: $t = 29.9$, $p < .001$]. It indicates the effectiveness of the songs in increasing the knowledge level of the farmers on importance of fodder production and other AH practices.

Fig. 2. CD of the songs developed

Fig. 3. CD being released by Hon'ble AH Minister, GoK

Songs were developed systematically using popular music tunes in Kannada language, and made available for extension work in Karnataka. They are being extensively used by the field veterinarians during animal health camps, training programs, etc. These traditional media outputs may be integrated with modern ICTs for effective dissemination of scientific animal husbandry knowledge.

Use of Skits for Effective Delivery of Animal Husbandry Extension Messages

Diversified use of teaching and learning methodologies by the field functionaries becomes at most important for effective delivery of Animal Husbandry (AH) messages. In this context, dramas and skit forms of the folk media have a lasting impact on the learning process. However, their use in AH extension is limited to university youth festivals and trainings of certain individuals and NGOs only. Lack of sensitization of the extension field functionaries in use of skits, lack of required skills and importantly lack of readymade scripts might be the reasons for being sparsely used in AH Extension. In this context, projects (2012 & 2022) were conceived with an objective of developing scripts on various concepts of AH and was funded by Directorate of Extension, KVAFSU, Bidar and was executed by Department of Veterinary and Animal Husbandry Extension Education, Veterinary College, Shivamogga and Hassan (Mooka Raaga).

A technique similar to Delphi was used to achieve the objectives. About twenty like- minded veterinary professionals were involved. The process involved identification of literary inclined veterinary professional; bringing them together under a single platform; sensitizing them to the technicalities of script writing for skits; encouraging them to pen the scripts; vetting the scripts; giving them the feedback and finalization of their scripts by the experts. It was carried out by conducting two workshops of one and two days respectively and an editorial committee meeting of one-day duration. In all, about 18 scripts on various concepts of Animal Husbandry were penned by the participants for Shivamogga college project and 11 scripts for Hassan college project.

Table 1: Details of the steps followed in the process of evolving scripts on AH concepts

S. No.	Particulars of the tasks
1.	Project formats were submitted to the Directorate of Extension, KVAFSU, Bidar for financial sanction and got the same. Meanwhile, Veterinary professionals across various institutions inclined in writing and interested to participate in the process were identified.
2.	Conducted workshop - I with the following objectives. I. To introduce and sensitize the participants to script writing II. To identify the AH concepts on which scripts can be prepared. III. Distribution of the concepts among the participants for evolving detailed scripts. The above objectives were achieved in due course of the process.
3.	The individual participants were encouraged to pen scripts on the concepts they chose at their work places.
4.	Scripts were collected from individual participants, typed and circulated among the editorial team members for vetting and discussion in the final workshop.

5.	Workshop - II was organized to give feedback to the authors.
6.	The authors incorporated the suggested changes, finalized the scripts and sent them to the coordinator.
7.	Editorial meet was convened to read the scripts thoroughly, made minor corrections and finalized them. Some scripts were even dropped.
8.	The final reports of the projects were submitted.
9.	University published the collection of scripts in the form of books.

Outcomes of the Projects

1. Networking of likeminded veterinarians interested in Veterinary science literature.
2. Sensitization of the participants to use alternative media like skits in extension work and were introduced to the technicalities of script writing for skits (especially, science based) to be used in AH extension.
3. A collection of 18 and 11 skit scripts was authored individually and jointly.
4. As a spillover effect, an organization, namely *Kannada Pashu Vaidya Sahitya Parishath* was initiated by likeminded Veterinarians involved in Kannada Veterinary Science Literature.

The documents were published by KVAFSU, and are being kept for sale. Further, 14 scripts were adopted for radio dramas by All India Radio, Kalaburgi. Interestingly, one of the script (Gadi datidaga) bagged second prize in the state level radio drama competition. Likewise, one of the scripts (Thabbaligalu) was adopted as a chapter in BA language text book by Mangalore University, Karnataka. The script of the skit Aashakirana - Women Empowerment through Animal Husbandry contested in MANAGE short Film competition 2023 and is linked to Youtube for public view (https://youtu.be/yqFbUPXYNNY). The script entitled Rabies was enacted in the form of skit by students of Veterinary College, Bengaluru to create awareness and educate people on preventive measures to be adopted in a dog bite condition in Mall of Asia, Bengaluru organized by BBMP on account of Rabies day in 2024. Likewise, the scripts from both the compilations are being used by students of various colleges in youth festivals and NSS programs. Selected few scripts can also be used in producing tele films, short videos which can form important extension tools.

Fig. 4. Short Film "Ashakirana" released by Hon'ble Vice chancellor and University officers of KVAFSU, Bidar

Fig. 5. Rabies skit by students at Mall of Asia, Bengaluru

List of concepts on which skits were developed

- Need for vaccination
- Need for boiling milk before consumption and careful handling of aborted material
- Misbelieves in animal husbandry
- Artificial insemination
- Rabies

- Proper feeding of cattle
- Role of small ruminants
- Care and management of backyard poultry
- Care and management of buffaloes
- Calf a year program
- Control of diseases
- Status of *desi* pig rearing
- Management of milk fever
- Veterinary profession as a career option
- Role of dairying in sustainable livelihood
- Importance of cattle insurance
- Importance of fodder production
- Enhancing profit in dairying
- Lumpy skin disease
- Calf rearing
- Importance of fodder
- SNF Problem
- Bloat
- Milk Fever
- Conception issues
- Home remedies for good animal health
- Plastic consumption issues in animals
- Retention of placenta
- Zoonotic diseases

Adoption of Government Schools for Capacity Building

KVAFSU also extended its outreach to schools under Prime Minsiter Modi's Vidyanjali II, Ministry of Education, Government of India Program, to inculcate awareness and interest in Animal Husbandry from an early age. This initiative comprises:

- Awareness programs to educate students on the importance of livestock in rural livelihoods.
- School farms, mini-dairies, college exposures to provide hands-on learning experiences.

- Workshops and interactive sessions to teach students about zoonotic diseases, pet care, and responsible animal ownership.
- Promotion of animal welfare ethics through student participation in animal care programs.
- Competitions and exhibitions to engage students in scientific learning.

Fig. 6. Awareness Program on "Rabies" and "Importance of Animal & Horticulture Produce in Child Nutrition" for school children of adopted school GHPS Haralalli, Hassan

II. Innovative Extension Approaches Under Collaborative / Sponsored Activities

The university is working in collaboration with many development departments and institutes in order to reach the farming community. It may be in the form of MOUs, Joint Ventures, Collaborative Programmes, Mission mode projects, etc. Some of the collaborative institutes in this regard are Indian Council of Agricultural Research, Government of Karnataka, Department of Science and Technology, NABARD, MANAGE, private companies, etc. Some of the important extension activities discussed in this paper are based on following projects and few such relevant experiences of the authors.

- World Bank funded KWDP-II, Sujala-III project of Government of Karnataka
- NAIP Component-III Project of ICAR, New Delhi
- Department of Science and Technology
- ICAR- National Agricultural Science Fund

Identification of Critical Gaps and Identified Interventions

Based on baseline survey results, conducted by project team during different projects, the production practices and critical gaps were identified in Animal Husbandry sector. Further, this identification of critical gaps has helped in framing suitable interventions for livestock development through these projects.

Table 1. Critical Gaps and Identified Interventions for Livestock Development

Dairy Production/Rearing System	
Critical Gaps	**Identified Interventions**
Low productivity	• Promoting balanced feeding • Increasing green fodder availability
Green fodder shortage	• Promoting High yielding fodder varieties cultivation • Introducing fodder trees like sesbnea and drumstick
Fodder wastage	• Promoting enrichment of dry fodder • Introducing chaff cutter • Promoting construction of Manger
Lengthy inter-calving period and delayed age at puberty	• Creating awareness and improving breeding facilities • Providing proper feeding and balanced ration • Organizing regular health and infertility camps
Others	• Promoting construction of Cattle sheds • Regular deworming based on faecal sample analysis • Training/Awareness programs
Goat/Sheep Production System	
Inbreeding	• Changing breeding ram/buck
Poor kidding & twinning rate	• Supplementary feeding during breeding and lambing/ kidding season/transition period
Kid and lamb mortality (20-25%)	• Deworming based on faecal sample analysis • Supplementary feeding during transition period
Unscrupulous marketing system	• Introduction of weighing machine
Others	• Training/Awareness program

To address the above mentioned critical gaps, technical interventions were introduced in consultation with the technical experts, field functionaries, farmers' groups and individual farmers. Some of the interventions addressed in the project area were:

- Awareness program to address the knowledge gaps
- Promotion of fodder production through establishment of fodder nurseries
- Establishment of rural hatcheries
- On-farm demonstration of Silage production through silo bags
- Introduction of improved breeds of bucks and rams to avoid inbreeding
- Demonstration on preparation of balanced ration
- Demonstration on fodder enrichment and fodder chaffing
- Supplementary feeding during transition period in goats
- Demonstration unit for Stall feeding of goats

- Diagnosis and control of subclinical mastitis
- Production of extension literature and documentaries
- Upgradation of local goats with Osmanabadi bucks in Bidar
- Supplementation of Urea Molasses Mineral Block (UMMB) and feed blocks
- Supplementation of mineral mixture
- Conservation of valuable local Deoni breed of cattle

KWDP-SUJALA-III Project in Karnataka

The project was implemented in selected 524 micro watersheds in two phases covering 2, 52, 687 ha (Phase I – 2910, MW – 140186 ha), Phase II 233 mw- 1, 12,501 ha) spread over nearly 163 gram panchayaths with an overall objective to demonstrate more effective watershed management through greater integration of programs related to rainfed agriculture and allied activities, innovative and science based approaches and strengthened institutions and capabilities of stakeholders at different levels.

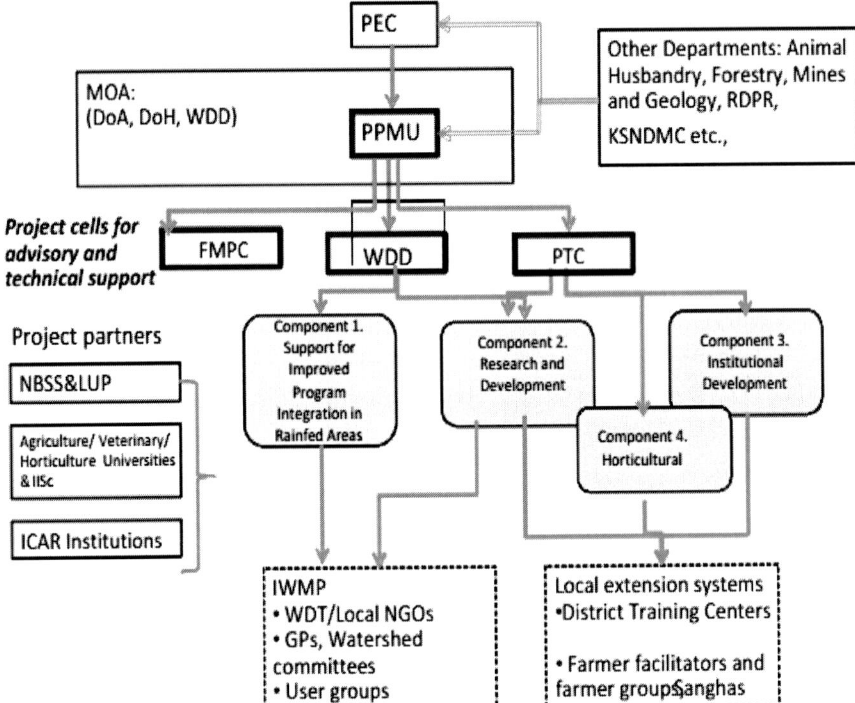

Fig. 7. Project Management Organogram- Linkages of Partners in KWDP-II (Sujala-III) project (Source: KWDP Report, 2017)

Awareness Programs/Workshops on Identified Interventions

The project was undertaken initially in seven districts, and later it was extended to 11 districts in which several awareness programs and on-farm demonstrations along with the line departments to the farmers on various need based and demand driven topics like green fodder production, dry fodder enrichment, silage making, supplementary feeding etc. were conducted. As per the target proposed by each district, about 6,000 farmers were benefited through these multiple programs in the project villages.

Prevention and Control of Sub-Clinical Mastitis

Demonstrations in all the villages of Sujala-III project districts about clean milk production, which includes disinfection and sanitation of shed, milking methods, cleaning of udder, strip cup test-mastitis detection and teat dipping were also conducted as a part of prevention and control of sub-clinical mastitis. At field-level Californian Mastitis Test (CMT) and Mastrip test were used to identify positive milk samples at pooled milk level, individual animal level and udder quarter level. Further, the positive samples were analyzed using Antibiotic Sensitivity Test (ABST) to determine the antibiotics suitable for treatment of the animals. Later, the positive animals were treated by the project staff with the help of Veterinarians at the earliest to prevent economic losses due to the disease. Further, follow-up of the treated animals was also conducted and the samples were screened again after the treatment, in second and third rounds as well.

About 7,800 samples were collected from all the seven Sujala districts, and analyzed it was found that 12% samples were positive. Further, a study conducted by Rathod *et al.* (2017a) indicated that economic losses due to Sub-Clinical Mastitis (SCM) were assessed to be in the range of INR 21,677 to INR 88,340 for one lactation period, depending on the type and condition of the animal. The study found that there was reduction in the occurrence of mastitis cases both in cows and buffaloes after imparting knowledge and skills on the mastitis detection and control techniques. The keeping quality of milk has also improved in both cow milk and buffalo milk. There was one lakh reduction in the somatic cell count of cows and buffaloes milk by the adoption of the practices. As a whole, this study has highlighted the effort of multidisciplinary team in prevention and control of SCM at field conditions and further, recommends that field problems have to be solved by participatory technology generation and transfer approach.

Fodder Production through Establishment of Fodder Nurseries in Project Districts

With many challenges at field level, the project teams in all the districts were successful in increasing the area under fodder production and fodder yield over a period of time. The establishment of these nurseries has focused on a combination of legume, non-legume and fodder trees in a limited area not affecting the area under food and cash crop production. Awareness programs, trainings, field days and demonstrations about fodder production and its importance were conducted by the multi-disciplinary teams for the beneficiaries. The study also involved fodder nursery and demonstration plot establishment, information dissemination through farm literature and video, regular follow-up by technical staff/human resource, access to inputs like fodder seeds, root slips, etc., and convergence of multi-stakeholders/ actors. A before-after research design was followed for the study to know the impact of these programs in the project villages.

The program has achieved the expected target, and has convincingly demonstrated increase in income generation, quality and quantity of milk yield, area under fodder cultivation and horizontal diffusion of fodder production practices (Rathod and Veeranna, 2017). A policy shift emphasizing delivery of inputs and regular follow-up for carrying out an integrated extension approach is very critical to enhance production and productivity.

Establishment of Rural Hatchery Unit

The constraints such as lack of availability of desi birds, lower hatchability, higher mortality and unhealthy chicks born were recognized. Hence, intervention was designed, where the farmers can be motivated towards organizing into groups adopting the concept of SHGs. This practice has also enhanced food and nutritional security and improved the supply and demand of poultry at local market. The eggs included Giriraja, Swarnadhara and desi birds in equal proportion. The total number of chicks pulled out was about 1,600 with average hatchability of 79.1 per cent. The total number of two weeks old chicks were sold @ Rs. 50/ chick by SHGs (Total = 1255) and total number of chicks distributed among the members were 220. Direct benefits to the farmers included:

- Availability of day old/ 4 weeks' desi chicks/improved birds at village level.
- Strengthened the rural food security through protein rich food of animal origin
- Supplementary income to the rural farmers by sale of birds and eggs

- Higher the hatchability of desi eggs
- Availability of assured quality chicks at village level

Supplementary Feeding of Goats during Transition Period

The program has achieved the expected target and convincingly demonstrated the importance of supplementary feeding of goats during transition period. There is an increase in income generation through improved kidding percentage, kid birth weight, kid growth, doe health condition, multiple births etc. Further, the project has observed decrease in economic losses due to kid mortality and problems during kidding. The project team also made efforts to motivate the beneficiaries, to strengthen their scientific knowledge and to make them adopt scientific practices in the form of supplementary feeding, deworming, vaccination etc.

There is an increase in income generation through improved kidding percentage, kid birth weight, kid growth, doe health condition, multiple births etc. Further, it was observed that the decrease in economic losses due to kid mortality and problems during kidding. The study has also witnessed the convergence of multi-stakeholders for achieving a common objective of improving productivity and hence, this experience of convergence can be applied at a larger scale.

Support to Downer Cows by means of Mobile Lift cum Trevis

One of the main difficulties faced in treatment of downer animals is task involved in lifting them up so as to stand again. While lifting the animal, lot of pressure, pain and injuries might occur to animal. More than all, such animals get frightened and discouraged to get up at all. In such situations, most of the times, farmers sell such animals to slaughter houses. If downer animals are made to stand comfortably, chances of their recovery will be good. Hence, the instrument "KVAFSU - Shiggaon Cow Lift" is designed. The instrument is so designed that with only one or two persons we can easily lift and make an animal weighing upto 800 Kgs to sit. Another attractive feature of the instrument is that; it can be carried to door step of the farmer. This machine has already begun functioning and has produced good results in the project area.

• NAIP Component- 3 Project

National Agricultural Innovation Project (NAIP) Component-3, Sub-project on "Livelihood Security through Resource and Entrepreneurship Management in Bidar District" was launched in Nov. 2008 at KVAFSU, Bidar. This project was implemented in 34 backward villages in four talukas (Aurad, Basavakalyan, Bhalki and Humnabad) of Bidar district, by a consortium of three partners with University of Agricultural Sciences, Raichur as the lead center. The subproject activities were initiated with Participatory Rural Appraisal to develop framework for suitable livelihood models for livelihood enhancement in 34 selected backward villages of Bidar district. Then initial project activities like benchmark survey, beneficiary cross verification and entry point activities were carried out. The output of these exercises helped in understanding the livelihood issues across the clusters and design appropriate interventions.

During initial period of project implementation, formation of Village Level Coordination Committees (VLCC) and livelihood groups received due attention. Based on the need, five different livelihood activity groups like integrated farming system, food processing, off farm activity, dairy and goat rearing, were formed by covering all the households during the project period. It includes small / marginal farmers, SC / ST and women-headed families, agricultural and landless labourers and dairy farmers. The project resulted in efficient utilization of available resources for increased production and income. It also played vital role in encouraging livelihood security through entrepreneurship development and establishment of economically feasible enterprises viz., small scale food processing units, upgradation of local goats. The project also bridged the existing gaps in improved technologies, linkages and developing self-management system to ensure sustainable livelihood security through organizing and strengthening 183 Community-based organizations. The project made following achievements :

- Increase in the cross bred population of Buffaloes and cows - 2171 (772 Buffaloes and 1,399 cattle) due to Artificial Insemination service at local level
- Conservation of Deoni breed of cattle
- Increase in milk yield from 504 ltrs to 850 ltrs/ animal/ lactation
- Focal breeder groups supplied 500 Osmanabadi bucks on cost basis to other goat rearers to upgrade 9600 goats of 372 households
- Increase in twinning rate by ten percent and kid birth weight by 0.65kg
- Establishment of dairy products preparation units
- For sustainability of the project 20 per cent contribution was collected for all the critical inputs which constitutes an amount of Rs.17,13,000. Out of this Rs. 11,50,000 kept as Fixed Deposit

Upgradation of local goats with Osmanabadi bucks in Bidar (Karnataka): In Bidar district, 480 Osmanabadi does and 240 Osmanabadi bucks were introduced to 24 livelihood groups at the rate of 2 does per member and one buck per group. As on date the focal breeder groups supplied 4800 Osmanabadi bucks to the other goat rearers in the project area to upgrade about 8,880 goats of 370 goat rearing households. It generated an income of Rs.1,35,000/- (including asset of goats) for the 240 local breeder group members and Rs. 40,000/- for the rest of the goat rearers (372 Households).

Conservation of Valuable Local Deoni Breed of Cattle: To conserve the valuable local Deoni breed of cattle and crossbreeding of non-descript cattle and dairy animals, four cluster level community managed AI centers were established in 2008-09 at each taluka in Bidar district to cater the services for six villages in each cluster (KVK, Bidar). Two unemployed youth from the identified villages in the cluster were trained in AI service for a duration of three months. The trained youth were initially asked to work under Assistant Director, Department of AH and VS for another three months. Later, they were allowed to work under VLCC as A.I. workers using the guidelines prepared. Semen straw, LN2 and other accessories were supported by the project, during the project period and later by VLCC/KMF on MoU basis. These four community managed cluster level A.I. centers are functioning independently since its inception. This has lead to establishment of 13 VMPCS's in the project area leading to increase in the milk production in the project area.

Supplementation of Mineral Mixture: Although the feed ingredients used in the preparation of the concentrate ration of dairy animals contain minerals, they are not sufficient to meet the needs of the animal. Hence, mineral mixture may be added to the concentrate ration of animals. Initially, the farmers were reluctant to add mineral mixture in the diet of their animals. Later on, when the farmers realized the importance of mineral mixture in the form of increased milk yield and improved body condition score, they started buying it in the packs of 5.00 and 10.0 kg. The mineral mixture was fed to adult dairy animals at the rate of 50-60g daily.

Supplementation of UMMB: Supplementation of UMMB, a brick shaped lick provides slow releasing urea nitrogen along with about 15 essential minerals. Uromin lick can supply 30-40% of protein and more than 75% of the daily mineral needs of medium producing dairy animals. It is an economical substitute of 1-2 kg concentrate mixture for grazing or stall fed animals. Its regular use can improve the reproductive performance of the cattle with better conception rate. The farmers were also trained to prepare the UMMB at their own farm. During different trials conducted in the area, it was observed that supplementation of UMMB to the lactating animals led to an increase in about 0.5 kg (average) milk yield daily besides improving their body condition.

Supplementation of Complete Feed Blocks: When fodder availability is limited, the best way to provide balanced ration to the milch animals throughout the year is by using fodder hay to make complete feed blocks. It is a compact feed, easy to transport and store. These feed blocks improve the utilization of crop residues; increases milk production and reduces the drudgery in animal feeding. The complete feed blocks were prepared by mixing wheat straw or paddy straw; concentrate mixture, molasses, urea, salt, mineral mixture and calcium oxide as per recommendations. Either five or 10 kg mixture was pressed into a block-making-machine and packed in plastic bags. It was fed directly to the animal after removing the plastic bag. The feeding of complete feed block has improved the dry matter intake, nutrient utilization and milk production in project areas.

Improving Fodder Availability through Green Fodder Production: The introduction of improved varieties of fodder crops through NAIP project has improved the health and production of livestock. Emphasis was given to multi-cut fodder varieties with a combination of legumes and non-legumes at field conditions. Before the sowing of these fodders, farmers were apprised on various recommended agronomic practices and were also contacted throughout crop season for improvement in the yield and quality of green fodder. By adopting the improved agronomic practices, green fodder availability was increased ultimately leading to improved productivity of livestock.

• Draught Mitigation through Feed and Fodder Management

Over the years, drought has been very intensive in Karnataka. In this context, the identified critical gaps related to fodder production and feeding are common across the different production systems - dairy, sheep and goat. Interventions, *viz*, establishment of fodder nurseries, silage making by using silo bags; enrichment of dry fodder and integration of fodder and fodder trees in horticulture crops and farm ponds may be carried out accordingly. The experiences of authors in this direction, though carried out in small scale and in isolation, have demonstrated the ways of how to mitigate fodder scarcity in summer season. From the accrued experiences, a five-pronged strategy is recommended for drought proofing small, marginal and other livestock farmers in dry land regions

- Emphasis on green fodder production in rainy season
- Green fodder grown in rainy reason is preserved as silage for summer season.
- Dry fodder available with the farmer is chaffed, enriched and used for animals
- Fodder trees shall be cultivated all along the bunds of the farmer's fields.

- Locally available non-conventional feed & fodder resources be utilized judiciously

Fig. 8. Five pronged strategy for Drought proofing Livestock Farming through Feed and Fodder Management

• Stakeholders and their Roles

Adoption of any technology is dependent on various factors such as inherent strength of the technologies themselves, their ease in adoption, availability of inputs, availability of technical knowhow, etc. In this context, various stakeholders involved in the process of development should be involved, empowered and they should perform their roles effectively.

#	Stakeholder	Role
1.	Farmer	• Gain knowledge and skill about the technologies • Procure inputs, adopt and use the technologies
2.	MPCS & KMF	• Facilitate in chaffing & supply of inputs • Arrange for trainings and demonstrations
3.	Dept. of AH&VS	• Facilitate in supply of inputs • Arrange for trainings and demonstrations
4.	Farmer Entrepreneurs	• Facilitate in chaffing of fodder on rent basis • Facilitate in silage preparation based on payment of service charges
5.	Input agencies	• Facilitate in supply of inputs like seeds, silo bags, drums, etc.
6.	Universities (KVAFSU)	• Conduct demonstration in all villages with more reach • Facilitate in training other stakeholders • Publicize the technologies using print, electronic, ICT media on a campaign mode

Group Approaches for Livestock Development

Group approaches in the form of SHGs, FIGs, LIGs etc. have been an effective way to achieve development through institutional mechanism. Majority of these groups take up livestock rearing as an income generating activity, which includes Dairy farming, Goat farming, Sheep farming etc. In the similar lines, Livestock Interest Groups (LIGs) were the groups promoted by the Department of Veterinary and Animal Husbandry Extension, Veterinary College Bidar, in similar guidelines to that of SHGs and integrated with goat rearing for the livelihood security of the members.

This goat rearing as an Income Generating Activity (IGA), which was supported and funded by DBT (Department of Biotechnology, Ministry of Science and Technology, Government of India, New Delhi) through project entitled "*An action research for self-employment of SC/ST youth through goat rearing under stall feeding as a strategy for poverty alleviation and sustainable development in Hyderabad-Karnataka region*". This project has highlighted the success of LIG model in Bidar district. Further, these concepts are also introduced in few other districts of Karnataka for other livelihood activities.

Development of e-Extension Modules

This project is designed with the objectives to develop e-extension modules in vernacular language (Kannada, Marathi, Telugu, Hindi and English) including e-content development in the field of animal husbandry for capacity building of the farming community and other stakeholders. The project is implemented by three Veterinary Universities of three states *viz.* Karnataka, Maharashtra and Telangana with KVAFSU, Bidar as the lead centre. The objectives include assessing the need for development of e-extension modules by multi-stakeholders in the field of animal husbandry; design and develop e-extension modules in local language for multi-stakeholders in the field of animal husbandry and Standardization and dissemination of e-extension modules on participatory approach. This project is expected to give access to credible and scientific information for all the stakeholders cost effectively in very less time, promote livestock health and enhanced quality of animal origin foods and conduct self paced learning modules and develop e-certificate courses.

Dairy Kannada App and Fodder Kannada App developed by KVAFSU are well accepted by the farming community of Karnataka and field-level professionals. A YouTube channel entitled "Pashu Pragna" has also been initiated and small concepts of veterinary and animal husbandry are being posted, using the platform to educate the farming community.

References*

1. KWDP Report (2017) Revised Project Implementation Plan of KWDP-II (Sujala III) Project. Watershed Development Department, Bengaluru, Karnataka. IDA Credit No. 5087-IN.
2. Rathod, P., Shivamurty V. and Desai A R. (2017a). Economic Losses due to Subclinical Mastitis in Dairy Animals: A Study in Bidar District of Karnataka. The Indian Journal of Veterinary Sciences and Biotechnology,13 (1): 37-41.
3. Rathod, P., Veeranna, K. C., Ramachandra B and Reddy D. (2017b) Utilization of Crop Residues for Livestock Feeding: A Field Experience. In: XXVI Annual Conference of Society of Animal Physiologists of India (SAPI) on "Physiological Innovations to Forecast the impact of Climate change and to evolve strategies for Sustainable Livestock Production" during 21-22 December, 2017 at Veterinary College, Bidar, Karnataka, pp: 66-71.
4. Rathod, P. and Veeranna, K. C. (2017) Innovative Approaches for Green Fodder Production and its Promotion in Dairy Development: Experiences of India. In: Fifth International Conference on Agriculture & Fisheries; Systems & Technology, during 08-09 December, 2017 at International Center for Research & Development, Colombo, Srilanka, pp:19.
5. Rathod, P., Veeranna, K. C., Ramachandra, B., Biradar, C., and Desai, A. R. (2018). Supplementary Feeding of Goats during Transition Period: A Participatory Action Research in North-Eastern Transition Zone of Karnataka State, India. International Journal of Livestock Research, 8(10), 238–246.
6. Shiva Rajkumar C M, K. Satyanarayan, L. Manjunatha, V. Jagadeeswary, Prakash R Rathod, S.B. Prasanna and B.N.V. Sowmya, February, 2025. Development of videos for social media platform (Youtube) for dissemination of animal husbandry technologies. Indian Journal of Veterinary & Animal Sciences Research, 54 (1) 48 – 55.

* The content of this chapter was prepared in consultation with above references

23

Extension Strategies of Dr. YSR Horticultural University, Andhra Pradesh to Reach the Unreached

E. Karuna Sree[1], A. Devivaraprasad Reddy[2], V Deepthi[2], K Mayuri[1], K Gopal[1] and B. Srinivasulu[1]

[1]Administrative Office, Dr.YSR Horticultural University, Venkataramannagudem, West Godavari District, Andhra Pradesh, India, [2]Krishi Vigyan Kendra, Dr.YSR Horticultural University, Venkataramannagudem, West Godavari District, Andhra Pradesh, India

Email: de@drysrhu.edu.in

Abstract

Digital extension strategies have become game-changing tools in the horticultural industry, providing scalable and inclusive ways to share knowledge and technology with farmers, especially those in remote and underserved areas. This paper looks at the many digital projects that Dr. YSR Horticultural University (Dr. YSRHU) in Andhra Pradesh has started to improve technology transfer and make horticulture more resilient. The study talks about new ideas that institutions have come up with, like the Farmers Advisory Cell (FAC), Plant Protection Advisory Cell (PPAC), Horticultural Skill Training Centre (HSTC), Horti Business Incubation Centre (HBIC), and the Virtual Farmers Training Centre (VTCC). These are all meant to help farmers get real-time advice, diagnostics, training, and business support. The focus is on using disruptive digital media, such as mobile apps like "Udyana Bandhu," WhatsApp and Telegram groups for specific crops, YouTube channels, community radio broadcasts, and phone-in shows to share information about crops and locations in local languages. Strategic partnerships with Rythu Bharosa Kendras (RBKs) make last-mile connectivity and feedback systems even more official. These actions have been shown to work in measurable ways to raise farmers' awareness, encourage the use of best practices, and give rural youth and women more power through skill development. This book chapter shows how ICT-based extension models are helping to close the digital divide and suggests that using these models on a national scale could be very important for achieving inclusive, long-term agricultural growth. In the end, combining digital strategies with traditional extension methods creates a hybrid model that can help improve knowledge systems and spark new ideas in Indian horticulture.

Keywords: *Digital Extension; Extension strategies; Outreach; Horticulture; ICT in Horticulture; Electronic wing; Community radio station*

Introduction

In Indian agriculture and its sub sectors contribute for about 18.4 per cent of the Gross Domestic Product (GDP). Globally, our country is being rated as on of the rapid growing major economies and rated as fifth largest economy by nominal GDP and in Purchasing Power Parity (PPP) its 3rd largest in the world. The horticultural crops have a special place in India's economy because they raise rural residents' incomes. India's horticulture production has surpassed food grain production and reached 320.11 million MT which is a huge accomplishment. In Andhra Pradesh the horticulture crops are spread across 17.84 lakh ha with 312.34 lakh MT yield. Andhra Pradesh is first in terms of fruit and spice producing area, and second in terms of micro irrigation area coverage. The horticulture industry makes for around 16.07 percent of the state Gross Value Addition. In India, Andhra Pradesh ranks first in the production of tomatoes, oil palm, papaya, cocoa, chillies, and cocoa, and second in cashew, mango, and sweet orange. Horticulture is increasingly being recognized as a sunrise industry because of its potential to boost farm income, provide a stable source of livelihood, and generate foreign exchange through exports. India's distinctive agro-climatic conditions, coupled with its vast diversity of crops and genetic resources, enable the country to produce a wide range of horticultural crops year-round.

The Government of Andhra Pradesh crafted Andhra Pradesh Horticultural University in Venkataramannagudem, West Godavari district in 2007, subsequently renamed as Dr. YSR Horticultural University in 2011. This was done in recognition of the importance of horticulture, its development potential in the state, and its contribution to the GDP. The university's mission is to develop human resources by teaching, doing research that is focused on local needs and conditions, and disseminating tested technologies through Extension operations.

The university is working on technology transfer (extension activities) through Krishi Vigyan Kendras (KVKs), Horticultural Research Stations (HRSs), Colleges of Horticulture, and Horticultural Polytechnics, as well as through Rythu Bharosa Kendrams (RBKs), District Resource Centers (DRCs), and the Department of Horticulture, Government of Andhra Pradesh. To "Reach the

Un-reached," extension is an ongoing effort that needs new ideas and time-limited methods.

Traditional agricultural extension has often been government-led and involves an extension officer visiting a farmer, a group of farmers, or holding farmer field schools. Nevertheless, despite its relatively long history and widespread adoption, providing extension services still faces significant obstacles, such as a lack of funding for public extension, a lack of participation by rural farmers and populations in extension activities, and a lack of research and effective extension methods. This restricts the availability of extension services, particularly in rural areas, and the ability to modify technical packages for local situations. The farmers and extension staff should benefit from these location-specific technologies as soon as feasible. This is crucial since extension workers at the village level lack easy access to the media, research institutions, and universities.

In order to bridge this gap in the delivery of extension and consulting services, a variety of cutting-edge methodologies have been created. In order to reach as many farmers as possible with extension messages, strategies such as farmer-to-farmer extension, farmer field schools, or farmer field days have been utilised. The emphasis on participatory learning and action, with more specialized services, such as facilitating access to markets and financial services, sets old extension programs apart from modern ones. The high expense of face-to-face extension, however, makes it difficult to provide the service to the farmers, who are frequently dispersed, in an efficient manner.

As an alternative to traditional face-to-face extension methods, Information and Communication Technology (ICT) enabled services which are being promoted by extension practitioners more and more. They span the internet, television shows, mobile technology services, and radio programs with add-on features. ICTs are thought to be inexpensive, and have the capacity to provide farmers—even those in remote places and with varied populations—with timely, pertinent, and actionable information. ICT-based solutions are also seen as a helpful tool for delivering extension services to women farmers in particular. Information given by ICTs is growing increasingly diverse, covering a range of topics such as specialised technology, market access, pricing formulation, meteorological data, and early illness, drought, and flood warnings. This enables farmers to decide what to cultivate and how to enhance their agricultural methods with greater knowledge. Rural residents in developing nations are increasingly likely to own a mobile phone, which creates an opportunity to use mobile-enabled extension messaging to provide smallholder farmers with the critical agricultural services they require.

E-extension or digital extension, will be given priority in order to address this issue. There are some successful initiatives in this approach. It is still in the early stages of need-based research despite the enormous need for knowledge. In order for scientists and extension authorities to immediately profit from the scientific knowledge, this information demand has to be further personalised using satellite information systems. The teleconferencing technique between the farmers and related agencies can be useful in resolving emerging issues in agriculture because adopting a new technology is essentially a decision-making process that necessitates analysis of the situation and resources, consultation, and frequent interaction with the farm scientists.

Dr. YSRHU Extension Strategies to Reach the Unreached

Extension strategies are very crucial for the rural development, socio-economic status and sustainable agriculture and allied sectors. The university is effectively reaching the unreached through better, faster, and effective way of approaching using the following extension strategies

Technology Assessment and Demonstration

In recent years, the agricultural landscape has been rapidly transforming, driven by advancements in technology. These developments have the potential to significantly improve productivity, sustainability, and profitability in agriculture and allied sectors. However, the success of these technologies depends largely on their assessment, adaptation, and demonstration to ensure they meet the needs of farmers and other stakeholders.

Understanding Technology Assessment

Technology can be understood as any organized body of knowledge and systematic actions that are applicable to routine activities. It involves applying scientific principles and knowledge for practical purposes, which allows humans to enhance their living conditions. Broadly speaking, technology encompasses both tangible and intangible entities that are created through the application of intellectual and physical efforts to achieve specific objectives. In the context of agriculture, the concept of technology can often lead to confusion among practitioners due to its complex nature, which integrates materials, processes, and knowledge. According to Swanson (1997), agricultural technologies can be categorized into two primary types:

1. Material technology, where knowledge is incorporated into a physical technological product; and
2. Knowledge-based technology, which includes the technical knowledge, management skills, and other processes that farmers require to improve farm management and support their livelihoods.

Technology assessment in agriculture involves evaluating new tools, techniques, and practices to determine their suitability, effectiveness, and impact. This process is crucial because not all technologies are applicable in every context. Factors such as local climate conditions, soil types, crop varieties, and socio-economic conditions can influence how well a technology performs. Therefore, thorough assessment helps in identifying the best-fit solutions for specific agricultural challenges.

The Role of Demonstration

Once a technology is assessed, demonstrating its use is the next critical step. Demonstrations are essential for showing farmers and other stakeholders how a new technology works in a real-world setting. By witnessing these technologies in action, farmers can better understand the practical benefits and challenges, helping them make informed decisions about adoption. Demonstration plots, field days, and farmer training programs are effective ways to showcase new innovations.

Training and Capacity Building Programmes

Training and capacity building are fundamental components for enhancing productivity, sustainability, and resilience in agriculture, horticulture, and allied sectors. These programs are designed to equip farmers, agricultural workers, extension officers, and other stakeholders with the necessary skills, knowledge, and competencies to adopt new technologies, improve practices, and effectively manage resources. In the context of evolving agricultural challenges and opportunities, the importance of these programs cannot be overstated. The training and capacity building programs enriches on the various topics includes, Enhancing Technical Skills and Knowledge**,** Promoting Sustainable Practices**,** Improving Farm Management and Entrepreneurship, Facilitating Adaptation to Climate Change, and Building Social Capital and Community Resilience.

Train the trainers (ToT) programs are specialized training initiatives designed to equip individuals with the skills, knowledge, and methodologies necessary to effectively train others. These programs focus on developing a cadre of skilled trainers who can deliver high-quality training sessions, ensuring the widespread dissemination of knowledge and best practices across various sectors. ToT programs are widely used in agriculture, horticulture, and other fields where capacity building is critical for development and success.

Farmer Field Schools (FFS) is a participatory approach where groups of farmers learn together through experimentation and observation in their own fields. This method promotes peer learning and helps farmers develop critical thinking and problem-solving skills.

Technical Support to RBKs

Rythu Bharosa Kendras (RBKs) are innovative, farmer-centric centers established in Andhra Pradesh, India, aimed at providing comprehensive support to the agricultural community. The initiative is part of the broader Rythu Bharosa scheme, which is designed to empower farmers by offering a range of services under one roof. The success and effectiveness of RBKs largely depend on the technical support they receive, which enables them to provide relevant, timely, and high-quality services to farmers.

RBKs are designed as one-stop service centers that cater to the various needs of farmers, including the supply of quality inputs, agricultural advisories, marketing support, and welfare scheme information. Each RBK is equipped with facilities to deliver a wide array of services, such as: Supply of Quality Inputs, Agricultural Advisory Services, Marketing and Price Information, and Testing and Certification.

Technical assistance is given to RBKs by scientists and educators working in several institutes under Dr. YSRHU through training sessions, in-person group talks with farmers and VHAs/VAAs of RBKs, as well as through webinars and Whats App groups.

Through the Integrated Call Center and RBK channel, digital platforms of the Government of Andhra Pradesh, scientists are also offering technical input to RBKs, technical assistance to RBK channel, and upgrading the most recent technology in horticulture and related areas.

Village Adoption through VC to Village Programme

The "VC to Village Programme," also known as "Managramam mana viswaviswadyalayam," is an innovative initiative aimed at fostering a strong connection between universities and rural communities. This program emphasizes the role of academic institutions in the holistic development of villages, focusing on sustainable development, education, technology transfer, and community empowerment. By adopting villages, universities take a proactive approach to addressing local challenges, utilizing their resources, expertise, and research capabilities to bring about positive change. With the motto to Transfer Knowledge and Capacity Building, Sustainable Development, Community Empowerment, Research and Innovation.

The VC to Village Programme encompasses several key components to achieve its objectives:

1. **Village Adoption**: Dr. YSRHU is having 42 institutes has to identify and adopt specific villages, committing to work with these communities over a period of one year. This long-term commitment allows for a deeper understanding of local issues and more impactful interventions.
2. **Capacity Building and Training**: The institutions under Dr. YSRHU regularly organizes need-based training sessions, workshops, and seminars to build the capacity of villagers. These training programs cover a wide range of topics, including modern agricultural practices, small-scale industries, digital literacy, health and hygiene, and financial management.
3. **Technology and Innovation Transfer**: The program facilitates the transfer of appropriate technologies from Dr. YSRHU to villages. This includes innovations in agriculture (e.g., improved seed varieties, irrigation techniques, improved package of practices, weed management), and Information and Communication Technology (ICT).
4. **Research and Development**: Faculty members and students engage in research projects that are relevant to the adopted villages. This research is aimed at solving local problems and is often carried out in collaboration with the villagers themselves, ensuring that the solutions are practical and acceptable.
5. **Animal Healthcare Initiatives**: The program includes animal health camps, vaccination drives, deworming, and awareness campaigns on nutrition, and scientific management practices, clean milk production etc.
6. **Monitoring and Evaluation**: Continuous monitoring and evaluation are integral to the program, ensuring that the interventions are effective and making necessary adjustments. Feedback from villagers is regularly collected to improve the implementation process.

Crop-specific Year-round Activities

Perennial crops like coconut, banana, and citrus require a consistent and systematic approach to management throughout the year to maintain healthy growth, optimize yields, and ensure high-quality produce. Unlike seasonal crops, these perennial crops have a continuous production cycle, necessitating year-round care. To focus on the development and revival of specific crops such as coconut, banana, citrus, and chili on a yearly basis. Dr. YSRHU has implemented a program one year one crop program to special focus on the above mentioned crops. In this program, offering certificate course, training programs, field visits, diagnostic visits, webinars, digital programs, etc on the improved varieties, package of practices, management aspects etc.

Field Diagnostic visits

Field diagnostic visits are critical components of modern agricultural practices, aimed at diagnosing and solving problems directly in the field. These visits involve experts, such as agricultural scientists, extension officers, and agronomists, who assess the health of crops, soil conditions, pest and disease presence, and other factors influencing agricultural productivity. The primary objective of field diagnostic visits is to provide on-site, real-time solutions to farmers, ensuring sustainable and optimal crop production.

Farmers Advisory Cell (FAC)

Agricultural extension and consultancy services are once again in demand as a result of agriculture's resurgence on the development agenda. Agricultural extension and consulting services make it easier for farmers, farmer groups, and value chain and market participants to transfer knowledge, information, better technology, and practices. The benefits of extension access on farmer knowledge, adoption, productivity, and financial returns have been demonstrated through research. A 58% return on investment is predicted for agricultural advice services. Established FAC in the university's main campus, serving as a coordination hub for 42 institutes working under Dr. YSRHU in terms of technology transfer. A mobile phone number (+91-9618021200) has been assigned to FAC particularly, and this number has become well-known among farmers for use in receiving frequent notifications and planning extension operations. For the benefit of farmers, VHAs/VAAs of RBKs, and field-level extension staff of the Department of Horticulture and Agriculture, Government of Andhra Pradesh, FAC routinely hosts webinars on technology in horticulture and related industries.

Plant Protection Advisory Cell (PPAC)

Sustainable agriculture crop production programs place a significant focus on plant protection strategies and practices. These plant protection initiatives aim to minimize agricultural losses due to pests by implementing integrated pest management, enforcing plant quarantine measures, regulating pesticide use, and providing locust warning and control. Additionally, these programs include training and capacity- building activities in plant protection to enhance farmers' skills and knowledge. PPAC was established at the university's main campus for frequent plant protection advisories as well as for monitoring and forecasting of pests and diseases. Regular alerts are provided in WhatsApp groups depending on the FPO.

Horticultural Skill Training Centre (HSTC)

The most important weapon for advancing the socioeconomic development of not only a person but also of the entire country is education. The provision of skill-centric learning and addressing the demands of business and society on a worldwide scale are two problems that higher education in the twenty-first century must overcome. The Horticultural Skill Training Centre at the University headquarters consistently offers skill training programs modules.

1. Dry Flower Technology
2. Advanced Dry Flower Technology
3. Processing & Value addition of fruits & vegetables
4. Advanced Nursery Management
5. Aquaculture Worker
6. Small Poultry Farmer
7. Terrace Kitchen Gardening
8. Scientific Beekeeping
9. Mushroom Cultivation
10. Production & Use of Biocontrol Agents

11. Value addition in jack fruit
12. Value addition in palmyrah
13. Cashew nut processing & its value addition
14. Mushroom production & value addition
15. Beekeeping
16. Production and use of organic manures

Horti Business Incubation Centre (HBIC)

You could be starting your firm with a limited budget, time, or staff, and you might be wondering how the business world truly works. You could wish you had more tools, a mentor, or more knowledge to support you as you expand your company and make it easier to sustain both the business and the cash. A business incubator may provide the tools your n ed to succeed if you're ready to expand your company but are unsure how to get beyond these obstacles. With the goal of assisting Horti/Agri students, rural youth, and women in becoming Horti entrepreneurs and establishing a "Employer Culture," the Horti Business Incubation Centre (HBIC) at the university's main campus offers services such as counselling, technical mentoring, business mentoring, capacity building, consulting, business plan preparation, incubation, a pilot scale production. An incubator is a company created to aid new businesses in growing and succeeding by offerin that is either free or inexpensive, mentoring, knowledge, access to investment in some circumstances, operating cost in the form of a loan. A college program is nearly exactly like entering an incubator in that one must apply, be accepted, and then adhere to a timetable in order to achieve the standards set by the incubator. To participate in the incubator, one need to make a time commitment, usually between one and two years.

Virtual Farmers Training Centre

Agricultural extension plays crucial role in building the rural and agricultural development. Extension efforts aim to influence people's behaviour in ways that have been predicted and are seen to be advantageous for both the individual and the broader community. For the purpose of transferring technologies to extension workers, farmers, rural youth, and women through online mode along with a facility for farmers and scientists to interact, the e-Extension initiative established the Virtual Farmers Training Centre at the university campus in Venkataramannagudem. Numerous programs were conducted for the benefit of farming community, rural youth, extension functionaries etc.

Dr. YSRHU Udyana Mitra

Smart phones usage is increasing day-by-day across the globe, and YouTube subscribers are increasing tremendously in a disruptive way. Seeing believes, all the viewers would like to have the content in digital media in the form of videos. In this direction, the Dr. YSRHU has established sophisticated electronic wing (audio and video wing) facility at university museum where routine recording of videos on latest technologies in Horticulture & allied sectors is being carried out by subject experts (Scientists/Teachers) under "Dr. YSRHU UdyanaMitra" Programme for dissemination of technologies & information to farmers, students, and stake holders as well as through Rythu Bharosa Kendra's (RBKs) channels for broader coverage.

Farmers Awareness Creation Centre (FACC)

Agricultural productivity of small and marginal holder is a key factor increase in rural economic and sustainable development. Small and marginal holder farmers frequently have limited access to information, which can be a significant barrier to productivity growth. The public agricultural extension service is one possible tool to lessen information limitation, although its efficacy has historically been modest. By lowering the cost of outreach efforts and assisting in better adjusting the information to the specific requirements and circumstances of farmers, digital technology might increase the efficacy of extension. Some farmers utilise digital extension services to get tailored advice on what crops to cultivate, what kinds and amounts of inputs to employ, and other cultivation techniques. Farmers Awareness Creation Centre (FACC) was established in the university museum and information on newly released varieties, cutting-edge crop protection techniques, and accurate diagnostic symptoms in a variety of horticulture crops were shown there. The FACC has an information kiosk set up as an electronic resource with the most recent technology that is accessible 24/7.

Community Radio Station

In addition to commercial and public broadcasting, community radio represents a third form of radio transmission. Community radio stations cater to specific geographical and interest-based communities, providing content that is relevant and valued by local audiences often overlooked by commercial or mainstream media outlets. A community radio station, named "Dr. YSRHU Udyana Vani," has been established at the institution's main campus, operating at a frequency of 90.8 FM. This station broadcasts daily programs covering weather forecasts, monthly updates on horticultural activities, and other relevant topics. This service is to share the information and also to create and contribute to media in a world where there is a wealth of media.

Dr. YSRHU Phone-in program

Every day, Dr. YSRHU's scientists and teachers are welcomed to the farmer's advisory cell (FAC) to make live telephone program as part of a program format known as a phone-in or call-in program. For improved debate and input from the phone callers, each day's phone-in program session will be entirely devoted to a single topic of general interest. The schedule of the phone-in program is released well in advance in newspapers, monthly magazines, and WhatsApp groups.

Transfer of Technology - Disruptive Social Media

India is home to a vast network of agricultural research institutions distributed throughout the country, encompassing various fields related to agriculture and allied sectors. Dr. YSRHU is having 42 institutions across Andhra Pradesh. Dr. YSRHU is working with a motto to reach the unreached in better, faster and cheaper way using innovative approaches using social media viz., YouTube channels, whatsapp groups, telegram groups, mobile applications, community radio station etc.

Dr. YSRHU YouTube Channels

A personal YouTube channel is available to everyone who joins YouTube as a member and similar to other social media sites. Created Dr. YSRHU has created YouTube channels and digital content is uploaded *viz.*, Udyana Varthalu, Udyana Vani, Udyana Mitra, Horti-news, Technical support to RBKs, University special events and visits of VIPs, webinars, virtual training programs, etc. The YouTube members who are interested in the channel are subscribing the channel. The viewers can watch the programs at their convenience.

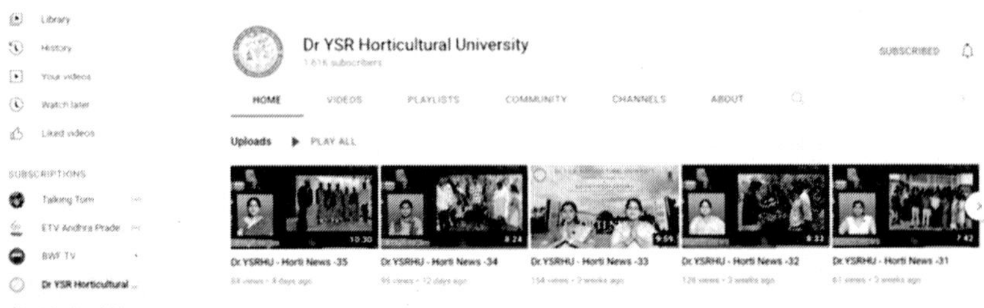

Crop-wise and Technology-wise WhatsApp groups

WhatsApp is free to use and one is not charged for any texts, photos, or video exchanges online. It is the modern texting app available to all smart phone users. The amount of storage on your phone does not matter in order to utilise

WhatsApp. The app is compatible with practically every smart phone and doesn't even require a high-end device. One may connect via phone number to WhatsApp so that others can use it to find there. This is more similar to the internet-based version of the messaging app on your phone. iOS, Android, and web- compatible devices may all use the program. The truth is that this program has grown to be really well-liked. Created about 50 crop-wise & technology –wise whatsapp groups with farmers, rural youth & women and providing regular advisories.

Dr. YSRHU Udyana Bandhu

There have been WhatsApp restrictions in recent years, and it became apparent that one cannot always rely just on one internet messenger. Fortunately, whether you have an Android or iOS smart phone, there are many of fantastic solutions available. One of the greatest applications in its category, Telegram, for instance, has amassed a sizable user base globally. At present, Telegram allows up to 2,00,000 individuals in groups that are established within the app, much beyond WhatsApp's current limit of 256 users, which is actually quite a lot and definitely more than you'll ever need. Although this may be an excessive number of users, it is still advantageous to WhatsApp over other apps to which it may apply. Many people even think it's a better alternative than WhatsApp because it has so many unique features and is so quick and simple to use. "Udyana Bandhu" telegram group was created to disseminate and interact with the officials of Department of Horticulture & VHAs/ VAAs of RBKs and the scientists of DrYSRHU for providing regular advisories, monthly calendar of operations, phone in programs, technical information etc.

Dr. YSRHU Udyana Bandhu app

Dr. YSRHU Udyana Bandhu application and made available as "Udyana Pantala e-Samacharam" in the google play store and works in both device (tablet and smartphone) is to provide advisories to telugu language farmers of Andhra Pradesh and Telangana. This mobile application facilitates data gathering and uploading as well as sharing information with farmers. An android app that serves as a mobile data information distribution tool is one component of the application. In this mobile application, tailored information covering about 33 horticultural crops under vegetables, fruit orchards, field orchards, flowers, medicinal plants and aromatic crops to farmers on the aspects of varieties, crop specific fertilizers recommendations, package of practices, integrated pest and disease management, post-harvest management etc. in local language for better understanding and communication.

ఉద్యాన పంటల e-సమాచారం

DRYSRHU

Webinars

Webinars have the capacity to engage a broad audience, allowing experts to share valuable knowledge, training, and expertise. They also foster relationships and build rapport by initially providing value, which in turn cultivates a loyal community and facilitates the transfer of technology. University under digital extension regularly conducting webinars on need based technologies in horticulture and allied sectors for the benefit of farmers, rural youth, women, VHAs/VAAs of RBKs and extension personnel of Department of Horticulture & Agriculture.

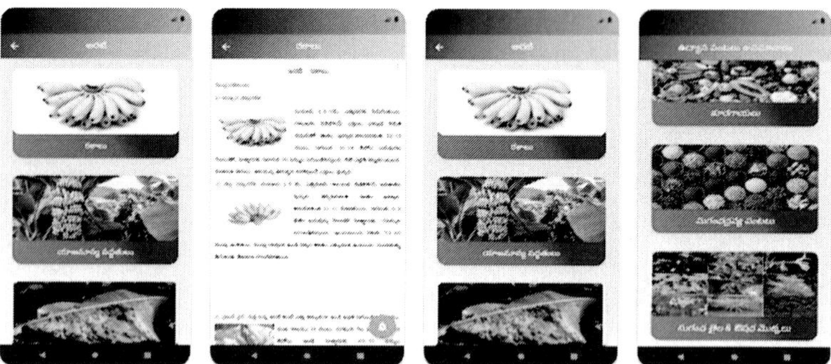

Pasidi Pantalu/ Phone-in-live /TV programs

Scientists of the university are participating in phone-in-live programs on latest technologies in horticulture in coordination with All India Radio and Doordarshan Kendra.

Dr.YSR HORTICULTURAL UNIVERSITY
EXTENSION SERVICES
Krishi Vigyan Kendras(4), Horticultural Research Stations (20), Colleges of Horticulture (8) & Horticulture Polytechnics (11)

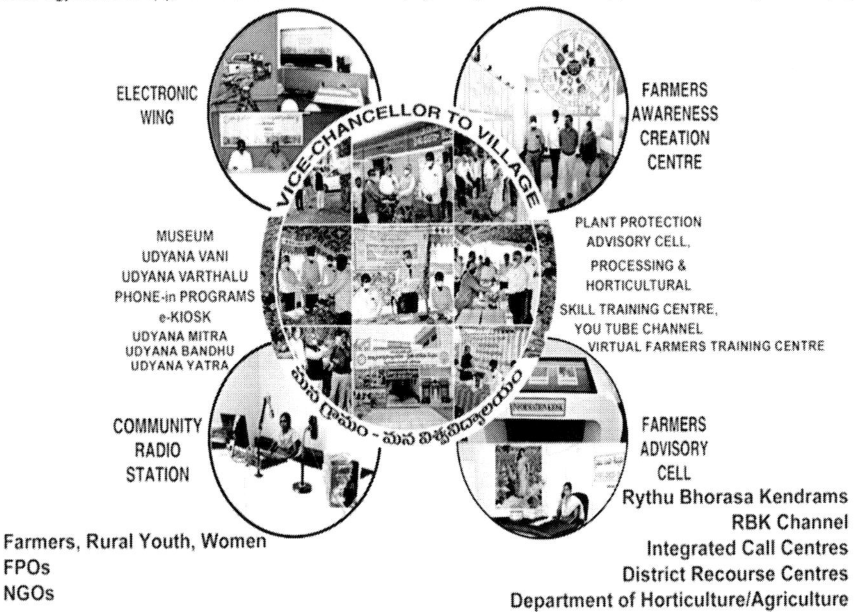

Conclusion

Population is increasing day by day and food is required for consumption. Extension strategies are very important for development of economic and sustainability. Traditional agricultural extension strategies have its own limitations and can able to reach few farming communities. ICT based innovative strategies are utilized to reach farmers in a better, faster, and cheaper product /services which can amplify the farmer productivity, nutrition, education, awareness on new technologies and results insustainable income.

References*

1. Alston, J., Chan-Kang, C., Marra, M., Pardey, P., & Wyatt, T. J. (2000). Research returns redux: A meta-analysis of the return to agricultural R&D. The Australian Journal of Agricultural and Resource Economics, 44(2), 185–215.
2. Amudavi, D. M., Khan, Z. R., Wanyama, J. M., Midega, C. A., Pittchar, J., &Nyangau, I. M. (2009). Assessment of technical efficiency of farmer teachers in the uptake and dissemination of push-pull technology in Western Kenya. Crop Protection, 28(11): 987-996.
3. Birkhaeuser, D., Evenson, R. E., &Feder, G. (1991). The economic impact of agricultural extension: A review. Economic Development and Cultural Change, 39(3),

607–650. Cai, T., & Abbott, E. (2013). Using video as a replacement or complement for the traditional lecture/demonstration method in agricultural training in rural Uganda. Journal of Applied Communication, 97(4), 1–15.

4. CASEY, M.J., MEIKLE, A., KERR, G.A. and STEVENS, D.R. (2016). Social media –a disruptive opportunity for science and extension in agriculture? Hill Country – Grassland Research and Practice Series 16: 53-60.
5. Dalwai, A. 2017.Empowering the Farmers through Extension and Knowledge Dissemination.Report of the Committee on Doubling Farmers' Income.Volume XI. Document prepared by the Committee on Doubling Farmers' Income, Department of Agriculture, Cooperation and Farmers' Welfare, Ministry of Agriculture & Farmers' Welfare.164 pg.
6. Davis, K. E., &Addom, B. K. (2000). Sub-Saharan Africa.In R. Saravanan (Ed.). ICTs for agricultural extension: Global experiments, innovations, and experiences. New Delhi: New India Publishing Agency.
7. Dercon, S., Gilligan, D. O., Hoddinott, J., &Woldehanna, T. (2009). The impact of agricultural extension and roads on poverty and consumption growth in fifteen Ethiopian villages. American Journal of Agricultural Economics, 91(4), 1007–1021.
8. FAO. Alternative approaches to organizing extension. (Source: https://www.fao.org/3/w5830e/w5830e04.htm)
9. FAO.Understanding extension. (Source:https://www.fao.org/3/t0060e/T0060E03.htm)
10. FAO.2021. Empowering smallholder farmers to access digital agricultural extension and advisory service. Food and Agriculture Organizations of the United Nations. (Source: Empowering smallholder farmers to access digital agricultural extension and advisory services (fao.org))
11. Jadoun, Y.S., Mukhopadhyay, C.S., Singh, A., Kaur, N. (2023). E-Agriculture Diaspora: Heralding A New Era of Animal Farming and Agricultural Practices. In: Mukhopadhyay, C.S., Choudhary, R.K., Panwar, H., Malik, Y.S. (eds) Biotechnological Interventions Augmenting Livestock Health and Production. Livestock Diseases and Management. Springer, Singapore. https://doi.org/10.1007/978-981-99-2209-3_24
12. Jayaraghavendra Rao V.K., R. Venkattakumar and Shahaji Phand (2021). Extension for Horticultural Technologies [e-book]. Hyderabad: National Institute of Agricultural Extension Management (MANAGE) and ICAR-Indian Institute of Horticultural Research. 71 Pg.
13. Kansiimea, M.K., Alawya, A., Allenb, C., Subharwalc, M., Jadhavd, A. &Parre, M. (2019) Effectiveness of mobile agri-advisory service extension model: Evidence from Direct2Farm program in India. World Development Perspectives, 13: 25- 33.
14. Monica K. Kansiimea, AbdillahiAlawya, Catherine Allenb, Manish Subharwalc, ArunJadhavd, Martin Parre. 2019. Effectiveness of mobile agri-advisory service extension model: Evidence from Direct2Farm program in India. World Development Perspectives, 13: 25-33.
15. Rajkhowa P, Qaim M (2021) Personalized digital extension services and agricultural performance: Evidence from smallholder farmers in India. PLoS ONE 16(10): e0259319. https://doi.org/10.1371/journal.pone.0259319.
16. Sajeev.M.V. & V. Venkatasubramanian (2010). Technology Application through KrishiVigyanKendras: Conceptual Paradigm, International Journal of Extension Education, INSEE, Nagpur. Vol.6:45-54. (14) (PDF) Technology Application, Refinement and Transfer through KVKs. Available from:https://www.researchgate.net/publication/329236901_Technology_Application_Refinement_and_Transfer_through_KVKs [accessed Sep 03 2024].

17. Swanson, B.E. (1997). Changing Paradigms in Technology Assessment and Transfer; Unpublished paper. Urbana, IL: University of Illinois, INTERPAKS.
18. World Bank. 2020. Digital solutions in a time of crisis. The World Bank Group, Washington, DC

* The content of this chapter was prepared in consultation with above references.

24

Frontline Extension Management Practices of Punjab Agricultural University

Manmeet Kaur[1], Gurpreet Singh Makkar[2] and Makhan Singh Bhullar[3]

[1]Associate Professor (Extension Education), [2]Senior Extension Scientist (Entomology), [3]Director of Extension Education, Directorate of Extension Education, Punjab Agricultural University, Ludhiana, Punjab, India

Email: directorext@pau.edu

Abstract

Punjab Agricultural University (PAU) is revolutionizing agricultural extension through a blend of digital innovations, stakeholder collaboration and capacity building. Its e-extension tools including the PAU Kisan App, Kheti Sandesh, social media platforms like YouTube and Facebook Live ensure timely and remote access to vital agricultural information. To strengthen engagement, PAU fosters direct farmer-scientist interactions through initiatives like Kisan Milani and the NRI Farmers' Conclave, while Farmer Information Centres facilitate grass root technology transfer.

PAU's robust Human Resource Development programs equip extension workers, farmers, women and youth with advanced skills in modern farming practices. The Research & Extension Specialists Workshops facilitate collaboration between research institutions and extension services, fostering a two-way communication channel to ensure that farmers receive relevant and practical knowledge. PAU conducts various training initiatives covering seasonal crop advisories, pest control techniques, water management strategies and specialized areas such as organic farming and precision agriculture. The Skill Development Centre and the Punjab Agribusiness Incubator (PABI) further promotes sustainable livelihoods and agri-entrepreneurship through hands-on training and startup support.

Together, these efforts in the domain of agricultural extension build an ecosystem that enhances productivity, profitability, sustainability, and resilience across Punjab.

India, a predominantly agrarian nation, relies heavily on agriculture for livelihoods. Over the decades, agricultural policies have focused on increasing the supply and accessibility of staple grains. This strategy has successfully reduced hunger and transformed the country's food system (Pingali *et al.*, 2019). At the forefront of this transformation stands Punjab that played a pivotal role in moving India from a food scarcity to self-sufficiency. Punjab's agricultural success is globally recognized, driven by the rapid technology adoption, enabled through synergy of infrastructure, incentives and most critically, information (Singh & Kohli, 1997).

Among these drivers, dissemination through agricultural extension proved instrumental in ensuring that innovations such as high-yielding variety (HYV) seeds, irrigation facilities and market incentives reached farmers effectively. While various factors contributed to Punjab's growth, but agricultural extension remains a cornerstone, translating research breakthroughs into field-level impact.

Punjab Agricultural University (PAU) has been a key catalyst in this journey. while serving as a premier institution in agricultural research, education, and extension. PAU's extension system has evolved in response to shifting agricultural needs, environmental challenges, and technological advancements. Beginning in 1960s, PAU led the Green Revolution by introducing HYVs and modern agricultural inputs through a dedicated, farmer-focused extension network (Gulati *et al.*, 2018).

In its early decades, PAU's extension services emphasized increasing productivity through mechanization and the use of fertilizers and pesticides. Initiatives such as Kisan Melas (Farmers' Fairs) and Kisan Clubs' facilitated knowledge transfer to rural communities, fostering large-scale adoption of new technologies. However, by the late 1980s, concerns such as soil degradation and pest resistance prompted a strategic pivot. The PAU approach shifted from purely productivity-focused methods to more sustainable and diversified practices such as Integrated Nutrient Management (INM), integrated weed management (IWM) Integrated Pest Management (IPM), water conservation, etc.

Today, PAU continues to modernize its extension practices through participatory, farmer-centric approaches. The current emphasis lies in promoting innovation, entrepreneurship, adaptability and sustainability through a blend of traditional outreach and modern ICT tools. Notable innovative approaches include e-extension services via mobile apps and online portals, real-time advisory systems through SMS and digital platforms, hybrid training modules combining both physical and virtual formats and the PAU Kisan App, which have modernized the way of information dissemination.

While physical outreach tools remain central to PAU's extension system, digital integration has expanded PAU's reach and relevance in an increasingly tech-driven rural economy. PAU's grass root level connectivity is further reinforced through its network of Krishi Vigyan Kendras (KVKs), beginning with its first KVK at Gurdaspur in 1982. Currently, the university operates 18 KVKs and 15 Farm Advisory Service Centres (FASCs), supported by a team of about 140 Extension Scientists. These centres are vital nodes for technology transfer, training and feedback loops, ensuring innovations are not only disseminated but also fine-tuned based on field realities.

Evolution of PAU Extension System

Punjab Agricultural University (PAU) has been the pioneer in transforming agricultural extension services in India. As the first institution in the country to introduce *Kisan Melas* (Farmers' fair) in 1967, PAU laid the groundwork for an outreach model that has significantly influenced the agricultural landscape of Punjab (PAU 2024).

Held twice a year in Ludhiana during March and September, these *Melas* quickly became a hallmark of PAU's extension efforts. These *Melas* feature field demonstrations, agro-industrial exhibitions, and farm produce competitions, attracting nearly 100,000 farmers from Punjab and neighbouring states such as Haryana, Uttar Pradesh, and Rajasthan. To reach farmers across diverse agroclimatic zones, regional *Kisan Melas* are also organized biannually before the *Rabi* and *Kharif* sowing seasons in sub-mountainous, rainfed and south-western regions.

During 1970s and 1980s, PAU played a pivotal role in turning Punjab into the "Breadbasket of India," focusing on high productivity in staple crops like wheat and rice. The introduction of the Training and Visit (T&V) System during this period aimed to standardize and streamline extension services (Feder and Slade 1986). While the T&V system increased the outreach, it faced criticism for being top-down and rigid (Tripathi *et al.*, 2023) and its overemphasis on external inputs like fertilizers and pesticides led to long-term environmental concerns. Recognizing these challenges, PAU initiated a strategic shift in the 1990s towards sustainable agriculture. The university's extension services began emphasizing crop diversification, IPM, INM and other environmentally sound practices to address the emerging environmental issues.

A key milestone was the establishment of the Agricultural Technology Information Centre (ATIC) in 1999 at PAU Ludhiana. As a single-window platform under the Directorate of Extension Education, ATIC offers farmers, access to latest technologies, advisory services and university-developed products (Singh and Kalra 2019). Located near the university's gate number one, it includes a farmer's information centre (plant clinic), a seed shop selling certified high-yielding varieties, vegetable kits, bio-fertilizers and a bookshop offering resources such as the monthly journals, *Progressive Farming* (English) and *Changi Kheti* (Punjabi), Package of Practices for *Kharif* and *Rabi* crops, Vegetables & Fruit crops, Crop Calendar, University Diary, and many other important publications. The early 2000s marked PAU's integration of Information and Communication Technologies (ICT) into extension services. Through radio, television, mobile advisory services and later, digital platforms, PAU significantly expanded its reach and responsiveness. These tools enabled real-time guidance on pest outbreaks, weather updates and crop management.

Since 2010, PAU has accelerated its digital reach. Mobile apps, online portals and social media now deliver instant access to expert recommendations, market trends and weather data. The Farm Advisory Service has become more efficient and farmer-centric, leveraging digital platforms to ensure timely, data-driven decision-making. ICT has proven particularly vital in rural areas, empowering farmers with continuous learning and actionable insights for sustainable agriculture.

Extension Approaches at PAU

Agricultural extension is not a one-size-fit all model, especially in Punjab, where farmers face diverse challenges and rapidly evolving farming practices. PAU has developed a multipronged, adaptable extension strategy, combining traditional methods with innovative, technology-driven solutions.

Key Approaches include:

General Extension & T& V System: Initially used to establish a strong outreach base and disseminate technical know-how efficiently.

Farming Systems Approach: Tailored to the holistic needs of farm households, integrating crops, livestock, and other farm enterprises.

Public-Private Partnership (PPP): Collaborations with different public and private sector stakeholders to broaden service delivery and resource mobilization.

Cyber Extension/ e-Extension: Leveraging ICT to deliver advisory services remotely via mobile phones, websites, and social media, breaking geographical barriers and enhancing accessibility.

Human Resource Development (HRD): Focused training for both extension personnel and farmers to build capacity in adopting modern techniques and tools.

Stakeholder Linkages: Strengthening collaborations with government bodies, research institutions and industry partners to enhance synergy and resource sharing.

PAU blends these approaches through a comprehensive extension framework that combines the strengths of legacy systems with innovative practices. This dynamic model ensures that farmers receive relevant, timely and data-driven support in an increasingly complex agricultural environment. The details of these approaches are as follows:

A. **e-Extension**
B. **Linkage Strengthening**
C. **Capacity Development**

A. e-Extension Approaches

e-Extension uses information and communication technology (ICT) to deliver agricultural extension services. It breaks geographical barriers, allowing farmers to continue learning and receiving guidance remotely, without the need for physical travel to distant locations. With a staggering extension worker-to-farmer ratio of 1:1000 (Kaur *et al.,* 2014), traditional methods fall short in reaching the vast farming population. Thus, e-extension approach is crucial in the current era, where digital connectivity can address the challenges posed by limited extension staff, vast rural areas and the increasing need for timely, updated information (Caria *et al.*, 2020). Ultimately, e-Extension reinforces the core mission of agricultural extension, empowering farmers with knowledge to boost productivity, ensure sustainability and adapt to dynamic agricultural environments.

PAU offers a robust e-Extension platform, where farmers can access information through mobile apps, websites, and digital magazines; attend online courses and virtual training sessions on various agricultural practices and participate in webinars on critical issues such as climate change, soil health and integrated pest management. These services allow farmers to continue learning and receiving guidance, even remotely. PAU actively engages with farmers through social media platforms such as Facebook, WhatsApp, and YouTube. Some of the major initiatives of PAU with respect to e-Extension are as PAU Kisan App, PAU website, PAU YouTube channel, PAU Facebook page, Digital magazine named *Kheti Sandesh* and virtual events described in details in further section:

Initiatives Related to e-Extension by PAU

1. PAU Kisan App

In the digital age, mobile applications have become a powerful tool in agricultural extension, transforming the way information is delivered to farmers (Barsbai *et al.*, 2020). These apps allow for the dissemination of real-time, location-specific advice, overcoming the traditional challenges of geographical distance and limited extension staff. By providing instant access to expert knowledge, mobile apps empower farmers to make timely, informed decisions, ultimately enhancing farm productivity and sustainability. These apps also facilitate quick information dissemination, which strengthens the extension process and makes it more usable for the farmers.

Fig 1. Interface of PAU Kisan app

To leverage the benefits of mobile apps in agriculture extension, PAU Kisan App was launched in Sept, 2019 to provide farmers with easy access to a wide range of agricultural information directly on their smart phones. More than 50,000 downloads were recorded within a short period of time. The app provides information about various aspects of agriculture and allied activities as given below:

- **Crop Information**: It provides detailed insights on crop varieties, sowing time, diseases, pests, and nutrient deficiencies.
- **Weather Updates**: It provides real-time weather forecasts to assist in planning agricultural activities.
- **Seed Database**: The app gives information on seed availability and quality for informed decision-making.
- **Training**: It provides information to training courses and workshops for skill enhancement.
- **Agricultural Advisories**: Personalized farming advice through the "*Kheti Sandesh*" is very useful feature of this app.
- **Multilingual Support**: The information can be obtained in Punjabi and English.
- **User-Friendly Interface**: The design of the app is very user-friendly for easy navigation.

So, the various type of advisories through PAU Kisan App viz. crop cultivation, production practices of different crops, pest management, agricultural machinery, crop residue management, weather advisory facilitates both farmers and extension personnel in effective outreach. Detailed information on training programs (district wise, month wise and subject wise) is also regularly updated on PAU Kisan App. Apart from this, digital newspaper named '*Kheti Sandesh*', district wise weather forecast and seed availability at different PAU centres are some of the prominent features of this app.

Regarding the content suitability of PAU Kisan app, Krishna (2023) conducted a study and revealed that a large number of users feel that the information regarding variety, weed management, insect pest management, disease management, fertilizer recommendation, irrigation, nutrient deficiency management and sowing methods were available on this app and the information was highly understandable and trustworthy. So, the app is expected to impact farmers in various positive ways, as with easy access to real-time weather updates, crop information and agricultural advisories, farmers can make informed decisions that enhance their productivity. Moreover, the app's information on pest management, disease control and nutrient management led to cost savings for farmers by reducing their reliance on external inputs. Ultimately, the app has immense potential to empower farmers with the tools and knowledge needed to improve their livelihoods and achieve sustainable agricultural practices.

2. PAU Website

In today's digital era, websites play a critical role in agricultural extension delivery, offering farmers and other stakeholders in agriculture, a centralized hub of information that can be accessed anytime and anywhere. These platforms allow for the rapid dissemination of extension literature and practical advisories, making the extension process more efficient and far-reaching. By providing open access to valuable agricultural resources, websites help bridge the gap between scientific knowledge and on-ground farming practices, promoting informed decision-making and innovation in agriculture. PAU's official website (www.pau.edu) is a prime example of how digital platforms can serve as a comprehensive resource for agricultural information.

Fig 2. Glimpses of PAU website

Features of PAU website related to extension services

- Hub of various ICT initiatives: It serves as a hub of all platforms as links of PAU YouTube channel, Facebook account, twitter handle, *Kheti Sandesh*, Farmer portal etc are provided on the website.
- Publications: The major publications of PAU viz. PAU monthly journals (Progressive farming & *Changi Kheti)*, Package of practices for *Rabi*, *Kharif*, Fruit, Vegetables and Flower crops are made freely available on the website both in Punjabi and English. Additionally, some other important literature to tackle the social and environmental issues are provided on the website to sensitize the masses. The catalogue of all publications of PAU can also be found on PAU website for ease of the farmers and other extension personnel.
- News and Updates: Regular updates on university news, events and achievements are posted to keep the visitors informed about various extension activities.
- Skill development: Farmers can fill google form provided on the website to be enrolled in various courses at Skill Development Centre, PAU.

Thus, farmers can easily find links to download the PAU apps and *Kheti Sandesh* visiting the website. Additionally, a dedicated portal is hosted on the website specifically for farmers, where they can access information on various aspects of farming, such as crop production, machinery, processing, subsidiary occupations, and publications. A separate section is designed to provide information tailored to farm women, promoting gender inclusivity in agricultural practices. Moreover, the website offers weather forecasts, seasonal crop advisories and disease alerts, helping farmers plan their activities accordingly. The major feature that benefits not only the farmers but field extension personnel also is the publications available on the website that are freely downloadable. The integration of PAU's website with other digital initiatives, such as mobile apps and e-extension tools, therefore, further strengthens the university's ability to reach a broader audience.

3. PAU Digital *Kheti Sandesh*

The shift from traditional print media to digital platforms has enabled agricultural institutions to disseminate information more efficiently and effectively, reaching a wider audience than ever before. Digital publications offer immediacy, accessibility, and the ability to engage with large number of farmers in short time. So, the digital versions of PAU literature and magazines like "*Changi Kheti*" and "Progressive farming" in the form of digital bulletin "*Kheti Sandesh*" have been developed to complement the traditional print versions. In the contemporary digital era, where a mere click could convey a message globally through mobile phones and internet-connected devices, PAU had launched its online digital magazine, *'Kheti Sandesh,'* in 2017. Released on every Wednesday, this digital magazine has successfully reached over four lakh farmers, providing timely and relevant information on a variety of agricultural topics.

Following are the features of PAU digital *Kheti Sandesh*

- Weekly Updates: The magazine is released every week, providing timely updates on agricultural practices, research findings and new technologies for the farmers.
- Farmer-Friendly: The content is designed to be easily understandable by farmers with practical advice and actionable insights.
- Wide Reach: It is distributed digitally through social media ensuring that farmers across Punjab and beyond can access the information easily.
- Personalized: Farmers' own success stories are integral part of Kheti Sandesh making it an effective tool for farmer-to-farmer extension.

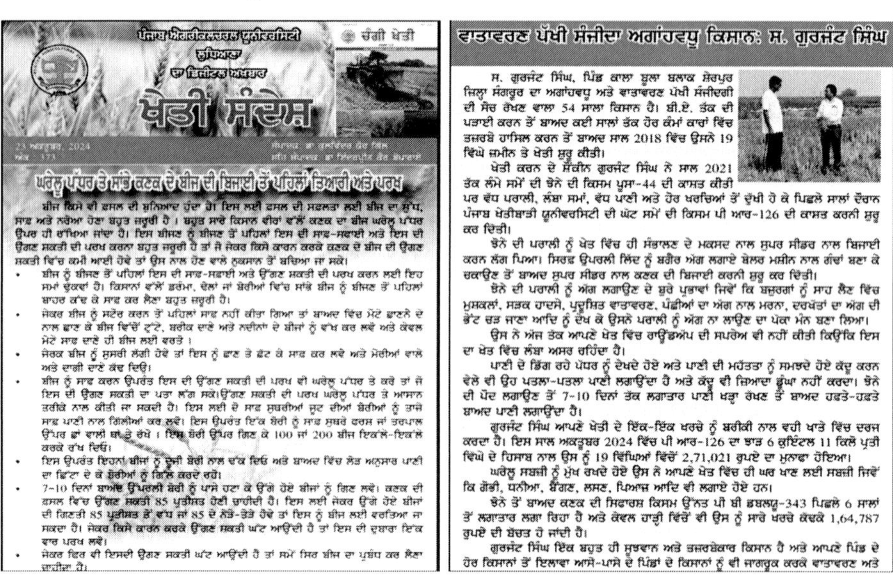

ਪੰਜਾਬ ਐਗਰੀਕਲਚਰਲ ਯੂਨੀਵਰਸਿਟੀ
ਲੁਧਿਆਣਾ
ਦਾ ਡਿਜੀਟਲ ਅਖ਼ਬਾਰ
ਖੇਤੀ ਸੰਦੇਸ਼

ਚੰਗੀ ਖੇਤੀ

ਘਰੇਲੂ ਪੱਧਰ ਤੇ ਸਾਂਭੇ ਕਣਕ ਦੇ ਬੀਜ ਦੀ ਬਿਜਾਈ ਤੋਂ ਪਹਿਲਾਂ ਤਿਆਰੀ ਅਤੇ ਪਰਖ

ਬੀਜ ਕਿਸੇ ਵੀ ਫਸਲ ਦੀ ਬੁਨਿਆਦ ਹੁੰਦਾ ਹੈ। ਇਸ ਲਈ ਫਸਲ ਦੀ ਸਫਲਤਾ ਲਈ ਬੀਜ ਦਾ ਸ਼ੁੱਧ, ਸਾਫ਼ ਅਤੇ ਨਰੋਆ ਹੋਣਾ ਬਹੁਤ ਜਰੂਰੀ ਹੈ। ਬਹੁਤ ਸਾਰੇ ਕਿਸਾਨ ਵੀਰਾਂ ਵੱਲੋਂ ਕਣਕ ਦਾ ਬੀਜ ਘਰੇਲੂ ਪੱਧਰ ਉਪਰ ਹੀ ਰੱਖਿਆ ਜਾਂਦਾ ਹੈ। ਇਸ ਬੀਜਣ ਨੂੰ ਬੀਜਣ ਤੋਂ ਪਹਿਲਾਂ ਇਸ ਦੀ ਸਾਫ਼-ਸਫਾਈ ਅਤੇ ਇਸ ਦੀ ਉਗਣ ਸ਼ਕਤੀ ਦੀ ਪਰਖ ਕਰਨਾ ਬਹੁਤ ਜਰੂਰੀ ਹੈ ਤਾਂ ਜੋ ਜੇਕਰ ਕਿਸੇ ਕਾਰਨ ਕਰਕੇ ਕਣਕ ਦੇ ਬੀਜ ਦੀ ਉਗਣ ਸ਼ਕਤੀ ਵਿੱਚ ਕਮੀ ਆਈ ਹੋਵੇ ਤਾਂ ਉਸ ਨਾਲ ਹੋਣ ਵਾਲੇ ਨੁਕਸਾਨ ਤੋਂ ਬਚਿਆ ਜਾ ਸਕੇ।

- ਬੀਜ ਨੂੰ ਬੀਜਣ ਤੋਂ ਪਹਿਲਾਂ ਇਸ ਦੀ ਸਾਫ਼-ਸਫ਼ਾਈ ਅਤੇ ਉੱਗਣ ਸ਼ਕਤੀ ਦੀ ਪਰਖ ਕਰਨ ਲਈ ਇਹ ਸਮਾਂ ਢੁਕਵਾਂ ਹੈ। ਕਿਸਾਨਾਂ ਵੱਲੋਂ ਡਰੰਮਾ, ਥੈਲਾਂ ਜਾਂ ਬੋਰੀਆਂ ਵਿੱਚ ਸਾਂਭੇ ਬੀਜ ਨੂੰ ਬੀਜਣ ਤੋਂ ਪਹਿਲਾਂ ਬਾਹਰ ਕੱਢ ਕੇ ਸਾਫ਼ ਕਰ ਲੈਣਾ ਬਹੁਤ ਜਰੂਰੀ ਹੈ।
- ਜੇਕਰ ਬੀਜ ਨੂੰ ਸਟੋਰ ਕਰਨ ਤੋਂ ਪਹਿਲਾਂ ਸਾਫ਼ ਨਹੀਂ ਕੀਤਾ ਗਿਆ ਤਾਂ ਬਾਅਦ ਵਿੱਚ ਮੋਟੇ ਛਾਣਨੇ ਦੇ ਨਾਲ ਛਾਣ ਕੇ ਬੀਜ ਵਿੱਚੋਂ ਟੁੱਟੇ, ਬਰੀਕ ਦਾਣੇ ਅਤੇ ਨਦੀਨਾਂ ਦੇ ਬੀਜਾਂ ਨੂੰ ਵੱਖ ਕਰ ਲਵੋ ਅਤੇ ਕੇਵਲ ਮੋਟੇ ਸਾਫ਼ ਦਾਣੇ ਹੀ ਬੀਜ ਲਈ ਵਰਤੋ।
- ਜੇਕਰ ਬੀਜ ਨੂੰ ਸੁਸਰੀ ਲੱਗੀ ਹੋਵੇ ਤਾਂ ਇਸ ਨੂੰ ਛਾਣ ਤੇ ਛੱਟ ਕੇ ਸਾਫ਼ ਕਰ ਲਵੋ ਅਤੇ ਮੋਰੀਆਂ ਵਾਲੇ ਅਤੇ ਦਾਗੀ ਦਾਣੇ ਕੱਢ ਦਿਓ।

ਵਾਤਾਵਰਣ ਪੱਖੀ ਸੰਜੀਦਾ ਅਗਾਂਹਵਧੂ ਕਿਸਾਨ: ਸ. ਗੁਰਜੰਟ ਸਿੰਘ

ਸ. ਗੁਰਜੰਟ ਸਿੰਘ, ਪਿੰਡ ਕਾਲਾ ਬੂਲਾ ਬਲਾਕ ਸ਼ੇਰਪੁਰ ਜਿਲ੍ਹਾ ਸੰਗਰੂਰ ਦਾ ਅਗਾਂਹਵਧੂ ਅਤੇ ਵਾਤਾਵਰਣ ਪੱਖੀ ਸੰਜੀਦਗੀ ਦੀ ਸੋਚ ਰੱਖਣ ਵਾਲਾ 54 ਸਾਲਾ ਕਿਸਾਨ ਹੈ। ਬੀ.ਏ. ਤੱਕ ਦੀ ਪੜਾਈ ਕਰਨ ਤੋਂ ਬਾਅਦ ਕਈ ਸਾਲਾਂ ਤੱਕ ਹੋਰ ਕੰਮਾਂ ਕਾਰਾਂ ਵਿੱਚ ਤਜਰਬੇ ਹਾਸਿਲ ਕਰਨ ਤੋਂ ਬਾਅਦ ਸਾਲ 2018 ਵਿੱਚ ਉਸਨੇ 19 ਵਿੱਘੇ ਜ਼ਮੀਨ ਤੇ ਖੇਤੀ ਸ਼ੁਰੂ ਕੀਤੀ।

ਖੇਤੀ ਕਰਨ ਦੇ ਸ਼ੌਕੀਨ ਗੁਰਜੰਟ ਸਿੰਘ ਨੇ ਸਾਲ 2021 ਤੱਕ ਲੰਮੇ ਸਮੇਂ ਦੀ ਝੋਨੇ ਦੀ ਕਿਸਮ ਪੂਸਾ-44 ਦੀ ਕਾਸ਼ਤ ਕੀਤੀ ਪਰ ਵੱਧ ਪਰਾਲੀ, ਲੰਬਾ ਸਮਾਂ, ਵੱਧ ਪਾਣੀ ਅਤੇ ਹੋਰ ਖਰਚਿਆਂ ਤੋਂ ਦੁੱਖੀ ਹੋ ਕੇ ਪਿਛਲੇ ਸਾਲਾਂ ਦੌਰਾਨ ਪੰਜਾਬ ਖੇਤੀਬਾੜੀ ਯੂਨੀਵਰਸਿਟੀ ਦੀ ਘੱਟ ਸਮੇਂ ਦੀ ਕਿਸਮ ਪੀ ਆਰ-126 ਦੀ ਕਾਸ਼ਤ ਕਰਨੀ ਸ਼ੁਰੂ ਕਰ ਦਿੱਤੀ।

ਝੋਨੇ ਦੀ ਪਰਾਲੀ ਨੂੰ ਖੇਤ ਵਿੱਚ ਹੀ ਸੰਭਾਲਣ ਦੇ ਮਕਸਦ ਨਾਲ ਸੁਪਰ ਸੀਡਰ ਨਾਲ ਬਿਜਾਈ ਕਰਨ ਲੱਗ ਪਿਆ। ਸਿਰਫ਼ ਉਪਰਲੀ ਲਿੰਦ ਨੂੰ ਬਗੈਰ ਅੱਗ ਲਗਾਏ ਬੇਲਰ ਮਸ਼ੀਨ ਨਾਲ ਗੰਢਾਂ ਬਣਾ ਕੇ ਚਕਾਉਣ ਤੋਂ ਬਾਅਦ ਸੁਪਰ ਸੀਡਰ ਨਾਲ ਕਣਕ ਦੀ ਬਿਜਾਈ ਕਰਨੀ ਸ਼ੁਰੂ ਕਰ ਦਿੱਤੀ।

ਝੋਨੇ ਦੀ ਪਰਾਲੀ ਨੂੰ ਅੱਗ ਲਗਾਉਣ ਦੇ ਬੁਰੇ ਪ੍ਰਭਾਵਾਂ ਜਿਵੇਂ ਕਿ ਬਜ਼ੁਰਗਾਂ ਨੂੰ ਸਾਹ ਲੈਣ ਵਿੱਚ ਮੁਸ਼ਕਲਾਂ, ਸੜਕ ਹਾਦਸੇ, ਪ੍ਰਦੂਸ਼ਿਤ ਵਾਤਾਵਰਣ, ਪੰਛੀਆਂ ਦਾ ਅੱਗ ਨਾਲ ਮਰਨਾ, ਦਰਖੱਤਾਂ ਦਾ ਅੱਗ ਦੀ ਭੇਂਟ ਚੜ ਜਾਣਾ ਆਦਿ ਨੂੰ ਦੇਖ ਕੇ ਉਸਨੇ ਪਰਾਲੀ ਨੂੰ ਅੱਗ ਨਾ ਲਾਉਣ ਦਾ ਪੱਕਾ ਮਨ ਬਣਾ ਲਿਆ।

ਉਸ ਨੇ ਅੱਜ ਤੱਕ ਆਪਣੇ ਖੇਤ ਵਿੱਚ ਰਾਊਂਡਅੱਪ ਦੀ ਸਪਰੇਅ ਵੀ ਨਹੀਂ ਕੀਤੀ ਕਿਉਂਕਿ ਇਸ ਦਾ ਖੇਤ ਵਿੱਚ ਲੰਬਾ ਅਸਰ ਰਹਿੰਦਾ ਹੈ।

Fig 3. PAU Kheti Sandesh

PAU *Kheti Sandesh* not only provide farmers with information on their mobile phone but also retain the true sense of newspaper by providing versatile information. Additionally, it disseminates information about scheduling of *Kisan melas* throughout Punjab and showcases success stories of various farmers who could serve as role models for their peer groups.

Regarding content and effectiveness of Kheti Sandesh, Singh (2024) conducted a study on the content analysis and reader's reaction on Kheti Sandesh and found that a large majority of farmers considered the content to be entirely practical with easy language, timely availability and were highly satisfied with their PAU digital magazine. Further it has been found to be very useful for field extension functionaries, as they can easily get the content required at a particular time and then disseminate information among the farmers. The study findings highlighted that majority of the extension functionaries found the content of the magazine to be very much understandable, highly relevant and practical and were highly satisfied with the information provided. Further these extension functionaries share their information with the farmers creating an exponential effect of the Kheti Sandesh.

4. PAU YouTube Channel

Video content has emerged as a powerful medium for knowledge transfer, particularly in the field of agricultural extension. YouTube serves as an effective platform for engaging diverse audiences, offering visual and auditory learning experiences that can simplify complex concepts. The ability to demonstrate practices and technologies in a relatable format enhances understanding and retention among viewers, making it a valuable resource for farmers and agricultural stakeholders, having lower cost and larger scalability (Abate *et al.*, 2023).

PAU's YouTube channel plays a pivotal role in this knowledge-sharing ecosystem. The PAU You Tube channel was launched in December 2015 and currently has more than 32 thousand subscribers including farmers, extension functionaries from KVKs (Krishi Vigyan Kendras) and stakeholders from the State Department of Agriculture and Farmers Welfare.

Fig 4. Snapshot of PAU YouTube channel

The PAU YouTube channel has various features as given below:

- Diverse content: PAU YouTube channel content is related to various aspects such as crop production, processing, marketing, motivational videos, farmer interviews etc. to cater the diverse needs of farmers, farmwomen, and other stakeholders. The channel provides a wealth of extension videos on various agricultural topics helping farmers stay informed about the latest practices and research.
- Expert Interviews: The channel includes interviews with agricultural experts and researchers, offering valuable insights and advice to the farmers.

Event recordings: It hosts recordings of *Kisan Mela* and other events conducted by PAU and make these resources accessible to a wider audience. **Interactive Elements**: PAU YouTube channel acts as a two-way extension tool where the viewers also engage with experts through comments, questions, or queries on the videos of the channel Thus, PAU YouTube channel helps the farmers by providing authentic information in this era of misinformation on digital platforms. The university posts regular updates, video tutorials and live webinars that cover various agricultural topics, such as demonstrations on the use of modern machinery and technologies and training videos on improved farming techniques. The university's YouTube channel hosts educational videos in Punjabi, making it an effective tool for knowledge transfer among the local community.

In a study conducted by Singh (2024) regarding the viewer responses and content analysis of the PAU YouTube channel, it was found that majority of the viewers consider the videos to be simple, easy to understand and motivating. Notably, the most significant engagement was observed regarding the scripted videos that includes scripted narratives and instructional videos on technology demonstration. This highlights the effectiveness of PAU YouTube channel in disseminating the information in an easy, understandable, cost effective and scalable way providing farmers with clear, actionable insights that they can implement on their farms.

5. PAU Facebook Live

Facebook serves as an effective tool for fostering real-time communication between agricultural experts and farmers due to its wide reach and interactive capabilities. Facebook is a very popular platform among the young as well as middle-aged farmers. This platform allows for live engagement, where farmers can pose questions, share experiences and receive immediate feedback from experts, creating a dynamic learning environment that extends beyond traditional extension methods.

PAU Facebook Live program was initiated with the aim to bridge the gap between farmers and agricultural scientists by providing a space for interactive discussions on various agricultural issues and reaching the farming community directly in real-time. These live sessions encourage direct dialogue and live interaction with the expert, thus enabling farmers to seek clarification on challenges they face in their farming practices.

The program features PAU scientists who share their expertise on a range of topics, including crop management, pest control, soil health, etc. The features of this program are as follows:

- Live Streaming: PAU conducts live sessions on Facebook, where experts discuss various agricultural topics, answer questions and provide real-time advice to farmers.
- Interactive Questions & Answer sessions: During the live sessions, viewers can ask questions and get immediate responses from the experts, making it a highly interactive experience.
- Variety of Content: The live sessions cover a wide range of topics, including crop management, pest control, new agricultural technologies, and best farming practices.
- Accessibility: Being on Facebook, it is easily accessible to farmers in wider areas and even remote locations.

- Recorded Sessions: Many of the live sessions are recorded and archived on the PAU Facebook page, allowing farmers to watch at their convenience.

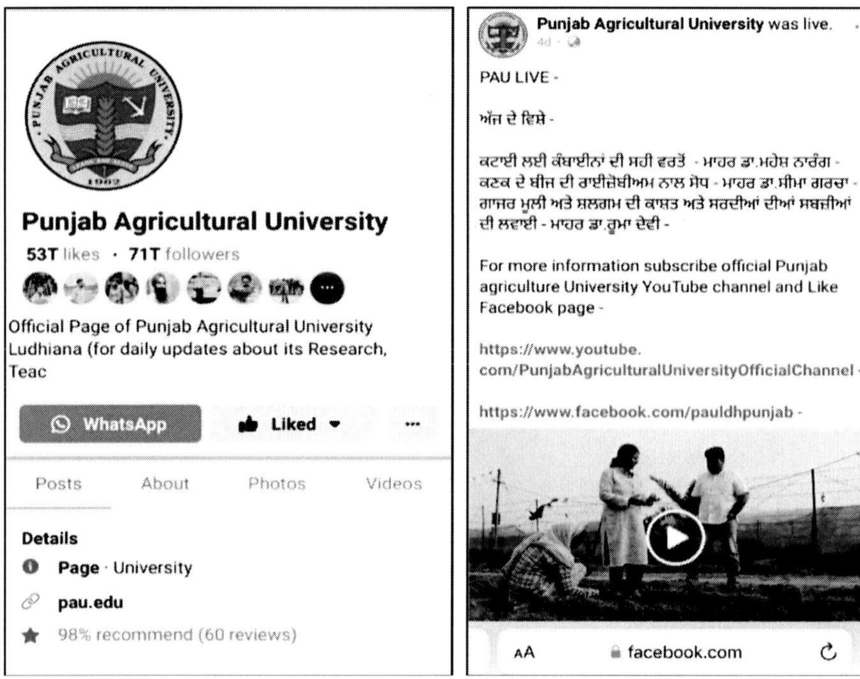

Fig 5. PAU Facebook live

By addressing farmers' specific concerns, the PAU Facebook Live program tailors the information to meet local needs, making it more relevant and actionable. Furthermore, the use of Facebook as a platform ensures that the reach of the program extends to a broader audience and serves as an accessible resource for farmers, especially in remote areas, allowing them to benefit from expert knowledge without the barriers of travel or scheduling conflicts. By using this platform, PAU not only disseminates knowledge but also empower farmers to engage actively in their learning process, leading to improved agricultural practices and greater resilience in the farming community.

6. Virtual Farmers' Training Programs and Kisan Melas

The rise of virtual platforms has revolutionized the way agricultural training and outreach is conducted, particularly in response to challenges such as the COVID-19 pandemic. These platforms such as google meet, zoom meet, etc offer innovative solutions for maintaining extension outreach ensuring that farmers have access to essential information and support, even in times of crisis. Virtual training programs not only provide convenience and flexibility but also allow for wider participation, breaking geographical barriers that often limit access to traditional extension services.

PAU has fully utilized this digital shift by conducting online training programs, *Kisan Melas* and workshops specifically designed for farmers. Notably, PAU was the first agricultural university to initiate a virtual *Kisan Mela* during the COVID-19 pandemic in 2022, demonstrating its commitment to supporting farmers during challenging times. Apart from virtual *Kisan Melas*, virtual Food & craft mela, farmer webinars, field days, farmer training camps and workshops were also started which continue till date to reach the farmers.

The features of these virtual programs, training programs, workshops and *Kisan Melas* are as below:

- Distant accessibility: Farmers can access the training programs and *melas* from anywhere eliminating the need of travel.
- Diverse topics: Due to online availability and cost-effectiveness with respect to travel, lodging & boarding, etc. farmers can take benefit of number of trainings on different topics without worrying about expenses.
- Whole-family approach: The benefits of virtual programs can be taken by the entire farmer family including youth and women.

The ability to attend training programs from home or even fields eliminated the barriers of travel and time, allowing more farmers to participate and gain knowledge that can significantly impact their farming practices.

Apart from embracing these digital technologies and e-extension, PAU continues to utilize other social media platforms such as WhatsApp and traditional media viz. television and radio. PAU collaborates with Doordarshan Kendra Jalandhar, local radio stations and private television channels to broadcast agricultural programs. These broadcasts and telecasts cover seasonal tips and expert discussions on crop cultivation, solutions for common agricultural challenges such as pest control and water conservation, success stories of progressive farmers who have adopted PAU's recommendations, etc. These media ensure that even farmers with limited internet access can benefit from expert guidance.

B. Linkage Strengthening Approaches

Building and strengthening linkages with stakeholders is essential to develop efficient extension ecosystem for fostering innovative technologies, enhancing productivity, and addressing the multifaceted challenges faced by the farmers. At PAU, the focus on strengthening these linkages encompasses collaboration with farmers, government agencies, research institutions and private sector partners. PAU has taken up various new initiatives for building and strengthening linkages with stakeholders through *Kisan Milanis*, NRI Farmer conclave, Farmer Information Centres, etc. as discussed below:

1. *Kisan Milani*

Kisan Milani a vital initiative of the Government of Punjab implemented through Directorate of Extension Education aimed at enhancing the efficiency of extension, research, and policy eco-system. It is essential in the current farming eco-system to obtain the first-hand feedback from the farmers and integrate this in research, extension, and policy eco-system. The Directorate of Extension Education, PAU has organized two *Kisan Milnis* in the year 2023 and one Kisan Milani in 2025. It was a unique initiative that brought together all the stakeholders viz. farmers, researchers, extension professionals and governance bodies from the entire state for a two-way exchange of information. These gatherings served as a forum for discussing the latest agricultural technologies, addressing farmers' concerns, and exchanging experiences in different agriculture sectors through panel discussions and interactive sessions amongst the stakeholders. The *Kisan Milani* initiative not only enhanced the dissemination of relevant information but also empowered farmers to actively participate in the decision-making processes that affect their livelihoods. These *Kisan Milanis* were organized at Punjab Agricultural University and apart from the researchers and extension personnel of PAU, extension personnel from all other government organizations such as Department of Agriculture and Farmers' Welfare, Department of Horticulture, Department of Soil and Water Conservation etc. were involved in the event to make the program comprehensive. Various parallel sessions where two-way interactions between the farmers and scientist/extension personnel were organized in the event. Further, feedback from the farmers was obtained from the farmers to form Agriculture Policy of Punjab.

Overall, this initiative embodies the philosophy of participatory development, recognizing that the path to agricultural advancement lies in the collective efforts and shared knowledge of all stakeholders involved.

2. NRI Farmers' Conclave

The role of NRI farmers in agricultural development is pivotal, as they often possess unique insights into global advanced farming techniques, market trends and investment opportunities that can significantly benefit the local agricultural landscape in Punjab. Through their experiences abroad, these farmers are well-equipped to introduce best practices in value addition, processing industries and other areas that are crucial for enhancing the economic viability of agriculture in the region.

PAU took a unique initiative by organizing NRI Farmers' Conclave aimed at creating a platform for dialogue, collaboration and investment in agricultural ecosystem of Punjab. By inviting NRIs to share their agricultural innovations and success stories, the conclave served as a catalyst for local farmers to adopt new technologies and practices that could increase their yields and profitability. Furthermore, the event emphasized the importance of value addition in agriculture and encouraged farmers in processing industries to transform raw agricultural products into high-value goods. This focus not only aimed to enhance the income of local farmers but also for employment generation and economic development in Punjab. This interaction created a platform where local farmers can learn directly from NRI farmers and gain practical knowledge that can lead to improved productivity, economic viability and sustainability of agriculture. By organizing NRI Farmers' conclave, PAU not only directed efforts for strengthening agricultural sector in Punjab but also created a model for sustainable growth that other regions can replicate.

3. Farmer Information Centres

To make the extension system effective, it is crucial to be present at grassroot level facilitating the active collaboration between scientists and farmers. The Department of Extension Education PAU, Ludhiana under the aegis of Directorate of Extension Education, has taken an innovative extension approach for effective delivery of technology, skills, and up-to-date information by developing the PAU Farmer Information Centre Extension Model. It serves as an intellectual and infrastructural framework within which the extension services are performed. The PAU-Farmer Information Centre (FIC) Extension Model was established in 2020 by the Department of Extension Education in Ludhiana district. This model of linkage emphasizes on the dissemination of information and technology transfer which takes place through various stakeholders like extension scientists, Agriculture Development Officers (ADOs), and secretaries of multipurpose cooperative societies. The extension scientists make regular visits to the adopted villages where these FICs are established in these adopted villages to cater to the needs of the farmers and other stakeholders in the entire circle. Apart from farm advisory, demonstrations are conducted for the PAU recommended technologies, promotion of PAU recommended varieties of various crops through seed availability at these centres, skill-based trainings for the farmers and exhibitions are being held in the villages operational under each Farmer Information Centre. The PAU-Farmer Information Centre Extension Model is framed around the linkages developed with the farmers through the liasioning bodies which are selected based on their extent of participation in the delivery mechanisms.

To determine the effectiveness of FICs, a study was conducted by Chauhan (2024) which revealed the farmers' satisfaction with services through FICs such as varietal promotion through seed distribution, promotion of recommended technologies etc. This feedback highlighted the importance of FICs in facilitating timely access to resources and enhancing the agricultural productivity of local farmers.

Apart from these new initiatives, PAU largely emphasize on training newly recruited scientists through induction trainings at research stations, arranging exposure visits for farmers and workshop, organizing conferences for researchers, extension functionaries and farmers. Induction training is an initiative by PAU where the newly recruited scientists are sent for induction training at various research stations to equip them with grassroot level understanding of the farm households and farming practices. PAU organizes exposure visits for farmers, extension personnel and other stakeholders to agricultural research institutions, model farms and progressive farms. These visits aim to help participants learn about successful agricultural practices, foster innovation, inspire the adoption of new farming technologies and create opportunities for farmers to see real-life applications of research in the field. PAU regularly holds *Kisan goshtis*, workshops, seminars, and conferences for different stakeholders within the extension ecosystem. These events focus on discussing the latest agricultural trends, challenges, and solutions, promoting dialogue between researchers, extension officers and farmers and sharing research findings and extension experiences across regions and disciplines. PAU has forged strong partnerships with industry, entrepreneurs, investors, and other stakeholders. They have successfully showcased branded products with attractive packaging, thus aligning with changing consumer preferences. PAU promotes commodity value chains and marketing strategies through collective bargaining with associations like the Progressive Beekeepers Association and PAU *Kisan* Club. Additionally, they encourage concepts such as on-farm markets, social media marketing, rural and urban *Kisan Bazars* and mobile van door-to-door sales.

C. Capacity Development Approaches

Punjab Agricultural University (PAU) focuses a comprehensive Human Resource Development (HRD) that plays a crucial role in strengthening the agricultural extension ecosystem. These programs aim to build the capacity of extension workers, farmers, and other stakeholders by equipping them with the latest agricultural knowledge, skills, practices, and behavioral changes. The HRD initiatives focus on improving the effectiveness of extension services, ensuring that scientific advancements in agriculture are efficiently transferred to the field level. PAU organizes a variety of training programs for farmers, government extension functionaries, scientists and other extension personnel working in the field of agriculture. Apart from regular seasonal trainings on latest crop advisories, pest control techniques and water management strategies; specialized training programs focused on evolving topics such as organic farming, precision agriculture, integrated pest management (IPM) and sustainable agriculture practices are also conducted.

1. Research & Extension Specialists Workshops

The concept of effective agricultural extension depends on collaboration between research institutions and extension service providers to ensure the dissemination of relevant and practical knowledge to farmers. Thus, regular Research & Extension Specialists Workshops for *Rabi* crops, *Kharif* crops and Horticulture crops are held at PAU to update the extension functionaries at KVKs and the State Department of Agriculture and Farmers' Welfare and further to obtain feedback from them regarding various technologies and recommendations at field level. About 350-400 officials from these departments attend these workshops regularly, which helps in strengthening the research extension collaboration in the state.

By systematically updating extension functionaries from the State Department of Agriculture and Farmers' Welfare and KVKs, the workshops ensure that the latest technologies and practices are effectively communicated to farmers. Additionally, these sessions provide an opportunity for extension functionaries to offer feedback on the relevance and applicability of various recommendations, fostering a two-way communication channel that enhances the effectiveness of agricultural extension efforts. This collaborative framework not only enhances the responsiveness of extension services to the needs of farmers but also promotes the adoption of innovative practices, ultimately leading to improved agricultural outcomes and rural livelihoods.

2. Skill Development Centre

With the vital role of skill development in promoting sustainable agricultural practices and enhancing rural livelihoods, the establishment of the Skill Development Centre at Punjab Agricultural University (PAU) represents a strategic response to the pressing need for skill enhancement in the agricultural sector.

Ministry of Skill Development and Entrepreneurship (MSDE) was formed in 2015 to enhance employability of the youth of Nation through Skill Development. Punjab Agricultural University got the honor to establish Punjab's first Skill Development Centre in agriculture with an aim to build capacity of its farmers, youth and women in agriculture as well as allied sectors and bridge the gap between laboratories and farm. Skill Development Centre established at Punjab Agricultural University has been affiliated as a training provider and Training of Trainers (TOT) program by the Agriculture Skill Council of India (ASCI) and Food Industry Capacity and Skill Initiative (SICSI). The memorandum of understanding (MoU) was signed with ASCI on September 10, 2015 and Food Industry Capacity & Skill Initiative (FICSI) on October 6, 2015 under the Prime Minister Kaushal Vikas Yojana (PMKVY) for various job roles. The Skill Development Centre focuses on the development of entrepreneurial skills to raise production and farm income as well as improve employment opportunity in agriculture and allied fields. Various courses ranging from a day to three months in the areas of production, processing, value addition, marketing, etc are organized to encourage the entrepreneurial skills of the farmers, youth, farm women and other stakeholders. The emphasis is given to equip the youth with knowledge, practical skills and its applications so that they can participate in nation's productivity and economic growth. The certificates of these courses are recognized throughout the country.

3. Punjab Agribusiness Incubator (PABI)

Entrepreneurship plays a pivotal role in driving economic growth, innovation, and sustainability within the agricultural sector. In today's era, Startups are vital for addressing contemporary challenges such as food security, climate change and the need for sustainable farming practices along with uplifting the livelihood. By fostering a culture of innovation and equipping individuals with the necessary skills and resources, entrepreneurship can significantly enhance productivity, create job opportunities, and improve the livelihoods of rural communities. So, recognising these needs, a new component under the revamped scheme *Rashtriya Krishi Vikas Yojana* (RKVY), Remunerative Approaches for Agriculture and Allied Sector Rejuvenation (RKVY-RAFTAAR) was launched in 2018-19. RKVY-RAFTAAR supports agribusiness incubation by tapping innovations and technologies for venture creation in agriculture. In this process, incubation facilities and expertise already available with participating academic, technical, management and R&D institutions in the country are utilized on an individual or collective basic to harness synergies. Punjab Agricultural University, Ludhiana has been bestowed with an incubator namely PABI under this scheme in March, 2019. There are two opportunities provided to the applicants under this incubator namely Uddam and Udaan.

Uddam, an opportunity to become an entrepreneur, is Agri-preneurship Orientation Programme aimed at providing an opportunity to students, youth, women and farmers or anyone having an innovative idea. The program offers Grant-in-aid opportunity for upto 5 Lakhs upon selection which will have to be used by the entrepreneurs to transform their innovative ideas to a prototype in the form of products / services/ processes/ business platforms or models etc. Udaan, an Opportunity to expand the business, is an Incubation Programme aimed at providing support to the Start-ups having innovative minimum viable products / services / business processes or models that are ready for commercialization. The program offers two-month workshop, mentorship, and Grant-in-aid opportunity for upto Rs. 25 Lakhs and facilitation support to the start-ups.

References

1. Abate, G. T., Bernard, T., Makhija, S., & Spielman, D. J. (2023). Accelerating technical change through ICT: Evidence from a video-mediated extension experiment in Ethiopia. *World Development, 161*, 106089.
2. Barsbai, T., Licuanan, V., Steinmayr, A., Tiongson, E., & Yang, D. (2020). Information and the acquisition of social network connections. *National Bureau of Economic Research*. Technical Report.
3. Caria, S., Franklin, S., & Witte, M. (2020). *Searching with Friends. Journal of Labour Economics, 44,4.*
4. Chauhan, P. (2024). *Performance analysis of PAU farmer information centre extension model* (M.Sc. thesis). Punjab Agricultural University, Ludhiana, India.
5. Feder, G., & Slade, R. (1986). The impact of agricultural extension: The training and visit system in India. *Research Observer, 1*, 139–161.
6. Gulati, A., Sharma, P., Samantara, A., &Terway, P. (2018). *Agricultural extension system in India: Review of current status, trends, and the way forward.*
7. Kaur, P., Kaur, K., & Kumar, P. (2014). Problem and prospects of privatization of extension services. *Research Journal of Social Science and Management, 3*, 89–94.
8. Krishna, U. (2023). *Content analysis of PAU Kisan App for major crops* (M.Sc. thesis). Punjab Agricultural University, Ludhiana, India.
9. PAU. (2024). *Kisan Melas*. Retrieved from https://www.pau.edu/index.php?_act=manageLink&DO=firstLink&intSubID=36
10. Pingali, P., Aiyar, A., Abraham, M., & Rahman, A. (2019). *Transforming food systems for a rising India* (Issue May). Palgrave Macmillan.
11. Singh, D. (2024). *Viewer response and content analysis of videos on YouTube channel of PAU Ludhiana* (M.Sc. thesis). Punjab Agricultural University, Ludhiana, India.
12. Singh, K. (2024). *Content analysis and reader's reaction on Kheti Sandesh: A weekly digital magazine of PAU Ludhiana* (M.Sc. thesis). Punjab Agricultural University, Ludhiana, India.
13. Singh, N., & Kohli, D. S. (1997). The Green Revolution in Punjab, India: The economics of technological change. *Journal of Punjab Studies, 12*(2), 285–306.
14. Tripathi, S., Gupta, B. K., Kumari, J., Pandey, A., & Mishra, B. P. (2023). *Advances in Agricultural Extension* (Vol. 1, pp. 121–131). Elite Publishing House.d Business Organizations (CBBOs)

Index